Advances in Fault Detection and Diagnosis Using Filtering Analysis

D1827293

Ziyun Wang · Yan Wang · Zhicheng Ji

Advances in Fault Detection and Diagnosis Using Filtering Analysis

 Springer

Ziyun Wang
School of IOT Engineering
Jiangnan University
Wuxi, Jiangsu, China

Yan Wang
School of IOT Engineering
Jiangnan University
Wuxi, Jiangsu, China

Zhicheng Ji
School of IOT Engineering
Jiangnan University
Wuxi, Jiangsu, China

ISBN 978-981-16-5961-4   ISBN 978-981-16-5959-1   (eBook)
https://doi.org/10.1007/978-981-16-5959-1

This Springer imprint is published by the registered company Springer Nature Singapore Pte Ltd.
The registered company address is: 152 Beach Road, #21-01/04 Gateway East, Singapore 189721,
Singapore

Preface I

At the core of many engineering problems is the solution of sets of equations and inequalities, and the optimization of cost functions. Unfortunately, except in special cases, such as when a set of equations is linear in its unknowns or when a convex cost function has to be minimized under convex constraints, the results obtained by conventional numerical methods are only local and cannot be guaranteed. This means, for example, that the actual global minimum of a cost function may not be reached, or that some global minimizers of this cost function may escape detection. By contrast, set-membership analysis makes it possible to obtain guaranteed approximations of the set of all the actual solutions of the problem being considered. This, together with the lack of books presenting set-membership techniques in such a way that they could become part of any engineering numerical tool kit, motivated the writing of this book.

There were at least two ideas on which we easily agreed, though. First, the book should be as simple and understandable as possible, which is why there are so many illustrations and examples. Secondly, readers willing to experiment with set-membership analysis on their own applications should be given the power to do so.

Wuxi, China

Ziyun Wang
Yan Wang
Zhicheng Ji

Preface II

Advances in Fault Detection and Diagnosis Using Filtering Analysis

In practical industrial applications, the operational safety and fault detection and diagnosis of engineering systems have received global attention. For actual engineering systems, how to find an efficient and accurate fault detection and diagnosis method to deal with the fault and ensure the safe and reliable operation of equipment is a very prominent problem in this field. To our knowledge, in the actual engineering system, there will inevitably be a variety of noises. In most cases, there is not enough data to summarize the random characteristics of these noises, and some noises do not have random characteristics, so it is difficult to describe them with statistical laws. Therefore, compared with traditional fault detection and diagnosis methods based on noises conforming to specific distribution laws, it is a meaningful work to conduct accurate and effective fault detection and diagnosis of systems with uncertain noises.

In all, this book proposes some filtering-based fault detection and diagnosis methods for engineering systems with unknown but bounded noises. In order to deal with the uncertain noises in the engineering system reasonably, this book uses the set-membership filtering method, which combines control discipline with space geometry. For different engineering systems, different spatial geometric shapes are used to contain the noises of the system, and the affine contraction process of the geometric space is used to describe the change of the state feasible set. Based on the obtained state feasible set, the process of fault detection and diagnosis can be further completed. On this basis, aiming at the problem of fault detection and diagnosis of engineering systems, from the perspective of filtering, this book not only improves the existing methods but also discusses and studies new methods. This book is of great value to post-graduate students, teachers, engineers, and individual researchers in the field of fault detection and diagnosis based on set-membership filtering.

Chapter 2 studies and analyzes the traditional fault diagnosis method based on filtering, and the Kalman filter algorithm is taken as an example to simulate and verify in the fault diagnosis of power converter. In Chap. 3, the fault detection method of set-membership filtering based on ellipsoid is analyzed and studied. Aiming at the problems of repeated calculation and low data utilization rate of the parameter estimation algorithm of weight ellipsoid, the parameter estimation algorithm of limited data window of weight ellipsoid is proposed. In Chap. 4, based on the polyhedral set-membership algorithm, a fault detection algorithm based on polyhedral set-membership filtering is proposed. In Chap. 5, based on the knowledge of interval operation and the idea of set inversion via interval analysis, a fault observer based on vector set inversion interval filtering is designed for the fault detection of DC motors. In Chap. 6, based on zonotopes and orthotopes, combined with linear programming theory, the fault diagnosis method of set-membership filtering based on polyhedron is studied. At the same time, considering the directional expansion theory and the stratification idea, the targeted research was carried out, respectively. In Chap. 7, two fault diagnosis methods based on composite set-membership filtering are proposed. Finally, in Chap. 8, the research contents of this book are summarized. Also, the future development direction of the fault diagnosis method based on set-membership filtering is prospected.

Wuxi, China   Ziyun Wang
June 2021   Yan Wang
                                                                                          Zhicheng Ji

Contents

Symbol Description

$\mathbb{R}$	Set of real numbers
$\mathbb{R}^n$	Set of n-dimensional real vectors
$\mathbf{B}^n$	Unitary box in $\mathbb{R}^n$
A^{T}	Transpose of matrix A
$\hat{A}$	Estimated value of matrix A
A^{-1}	Inverse of matrix A
$det(A)$	Determinant of matrix A
$tr(A)$	Trace of matrix A
$A > 0$	General notation for strictly positive definite matrix A
$A \geqslant 0$	General notation for positive definite matrix A
$A < 0$	General notation for strictly negative definite matrix A
$A \leqslant 0$	General notation for negative definite matrix A
$\mathbf{I}_n$	Identity matrix in $\mathbb{R}^{n \times n}$
$\mathbf{0}_n$	Zeros matrix in $\mathbb{R}^{n \times n}$
$\mathrm{diag}(a_1, \ldots, a_n)$	Diagonal matrix of dimension n
$\lvert \cdot \rvert$	Absolute value
$\lVert \cdot \rVert_\infty$	Infinity norm
$\lVert \cdot \rVert_P$	P-norm
$\lVert \cdot \rVert_F$	Frobenius norm
$s.t.$	Subject to
$\in$	Belongs to
$\notin$	Not belongs to
$\subset$	Subset
$\supset$	Contained in
$\cap$	Intersection
$\oplus$	Minkowski sum
$\odot$	Linear mapping
$conv(\cdot)$	Convex hull
max	Maximum value

| min | Minimum value |
| Ø | Empty set |

List of Figures

List of Tables

Chapter 1
Introduction

1.1 Fault Detection and Diagnosis Problem

With the further development of industrialization, the scale of electric power [1], chemical industry [2], machinery [3] and other industries has been expanding, and the system structure has become more and more complex. Affected by many unpredictable and unavoidable factors, these large industrial systems may have various faults at any time. Therefore, fault detection and diagnosis plays a key role in the operation and maintenance of equipment. In the process of industrial production, the synchronous guarantee of production efficiency and system safety has always been the focus of people's attention. Once the system components or subsystems fail, the system usually can not operate as expected. In this case, if the system continues to run, it will bring significant losses to the economic benefits, equipment maintenance, and the employee safety [4, 5].

As a renewable energy source, wind energy has the advantages of large energy storage, pollution-free, sustainable, and renewable energy. It is one of the most economical green energy sources in the world [6]. Wind turbine is a complex electromechanical system responsible for converting wind energy into electric energy, and it is generally composed of system components such as hubs, transmission shafts, gearboxes, generators and so on [7]. Wind turbines are exposed to the environment of large temperature differences between day and night, big load change and random wind impact all year round. The working environment is very harsh, and they are prone to failure. At the same time, the remote industrial environment of the wind turbine is remote, faults are usually difficult to find and repair in the first time, and it is easy to evolve into major accidents, which greatly increases the maintenance cost [8, 9]. It is reported that in 2004, Danish motor suppliers lost 40 million Euro due to the motor faults of wind turbines. Power converter is a widely used power conversion device in system energy conversion and transmission, and it plays a key role in power, industry, agriculture and other fields [10]. With the expansion of the application field of power electronic technology, the application field of power converters will be broader. However, as an intermediate link of power system conversion,

Z. Wang et al., *Advances in Fault Detection and Diagnosis Using Filtering Analysis*, https://doi.org/10.1007/978-981-16-5959-1_1

power converter is vulnerable to higher switching frequencies. At the same time, its work is also accompanied by electric heating, which leads frequent faults of power converter components [11, 12]. In addition, the safety and reliability of industrial products such as air conditioners, electric vehicles and elevators that are commonly used in daily life are also issues that cannot be ignored.

Fault detection and diagnosis play a key role in the operation and maintenance of equipment in these industries. Regular maintenance of the system is usually more economical and safer than dealing with faults after they occur. When the system fails, the performance of the components or subsystems of the system cannot meet the requirements, which will lead to the increase of resource consumption, the decrease of system performance or the loss of its intended functions, or damaging the mechanical equipment, causing the entire system to paralyze and resulting in a huge economy loss in the worst case, and even endangering the personal safety of employees. Therefore, fault detection and diagnosis is also important in cost management, efficiency improvement and environmental protection. The development of science and technology has promoted the progress of industrialization, and the strict requirements on industrial safety has been put forward while the demand for product quality and production efficiency in various industries has further improved. Therefore, using safe and reliable fault detection and diagnosis methods for timely and effective fault detection and diagnosis of the system has important practical significance to ensure the reliable performance of the system, improve the economic benefits of the system, and avoid casualties and environmental pollution.

1.2 Classification of Fault Detection and Diagnosis Methods

The research of fault diagnosis technology originated from Beard's doctoral dissertation published in 1971 [13]. With the investment of a large amount of capital, manpower and material resources, fault diagnosis technology gradually developed. Based on the researched contents of fault detection and diagnosis, Willsky published the first review article related to fault detection and diagnosis in Automatica in 1975 [14]. Subsequently, the first academic work in the field of fault detection and diagnosis was published in 1978, laying a solid theoretical foundation for the development of subsequent fault detection and diagnosis technology [15]. With the research and discussion of scholars for nearly half a century, fault diagnosis technology has shown a good development trend along the way. Various fault diagnosis algorithms have been proposed, developed, improved and matured, which can effectively deal with various systems and various situations. Under the fault diagnosis problem, and at the same time have good diagnosis accuracy and speed. At present, fault diagnosis methods are mainly divided into three categories: analytical model-based method [16–18], knowledge-based method [19, 20] and signal-processing-based method [21–23].

Furthermore, numerous experts and scholars have applied these methods to fields of real industries. For example, Qiu and Dai [17] proposed a chemical process fault diagnosis model and applied it to the Tennessee Eastman process. Jiang et al. [18]

developed a model-based fault diagnosis method for drill string washout that uses an iterated unscented Kalman filter to detect the emergence of drill string washout and to estimate washout depth and rate.

1.2.1 Analytical Model-Based Method

The analytical model-based fault diagnosis method is the earliest research methods, and is also the most in-depth and mature method. This method needs to be based on the precise mathematical model of the system to be diagnosed. The system is diagnosed by the method according to certain mathematical means, which showing high diagnostic accuracy. The method is not suitable for systems which is difficult to obtain an accurate mathematical model, and it is sensitive to unknown faults at the same time. The analytical model-based fault diagnosis method can be divided into state estimation method, parameter estimation method and equivalent space method. There are certain connections between the three methods, but they develop independently at the same time. Both the state estimation method and the parameter estimation method belong to the field of estimation [24, 25]. The accuracy of fault diagnosis depends on the accurate analysis of the system model, that is, by processing a series of actual observation data with interference signals and measurement noise, we can obtain various estimated state or parameter values. The basic principle of the equivalent space method is to detect the equivalence of the system mathematical model of the diagnosed object through the actual system input and output values, so as to carry out fault diagnosis. That is to say, according to the relationship between the input and output, the direct redundancy or instantaneous redundancy of the diagnosed system is given, and then the fault detection result of the system can be obtained [26].

The analytical model-based fault diagnosis method is widely used in power systems [27], chemical processes [28], vehicle systems [29], etc. In order to solve the problem of fault diagnosis and fault tolerance of linear drive system under the influence of system noise, Huang et al. [30] proposed a fault detection method based on Kalman filter residual generator. Then, two Kalman filters are designed to judge the fault type of system. Finally, when the fault is diagnosed, fault-tolerant control is used for fault adaptation. And the effectiveness and feasibility of the proposed algorithm are verified by the test and simulated in the real linear drive system. In Liu and He [31], for the sensor fault diagnosis problem of series battery pack, an adaptive extended Kalman filter is designed to estimate the state of each single battery. Then the results of fault detection and isolation can be obtained based on the residual error evaluation between the estimated value and the measured value. And finally the effectiveness of the proposed sensor fault detection and isolation method is verified under the UDDS driving cycles. In order to solve the multiple fault diagnosis problem as well as the sequence of all the potential multiple faults simultaneously, Lv et al. [32] applied a novel multiple fault diagnosis method based on the dependency model method, which uses the knowledge to test results and the prior probability of each fault type. The numerical example and case study verified the effectiveness and superiority of the proposed method.

1.2.2 Knowledge-Based Method

The knowledge-based fault diagnosis method does not require an accurate system mathematical model, nor does it need to have a deep understanding of the specific working mechanism of the system. It is a method to process the object information and realize the fault diagnosis in a knowledgeable and intelligent way from the perspective of knowledge [33, 34]. With the development of artificial intelligence technology, the application research of knowledge-based fault diagnosis method is deepening and expanding. This method is suitable for the field of nonlinear systems, skillfully uses the characteristics of intelligent knowledge, and has a good research prospect. However, at the same time, it also has the characteristics of being too dependent on relevant experience knowledge. Knowledge-based fault diagnosis methods mainly include expert system-based method, neural network-based method, and knowledge-based fault tree method.

In order to detect the operation and fault status of induction motor as early as possible, a fault diagnosis system for induction motor based on convolution neural network model is developed in literature [35]. By inputting the vibration signal data obtained in the experimental environment into the convolution neural network, the system fault can be judged. The induction motor system performs three fault diagnosis states: normal, rotor fault and bearing fault. The experimental result verifies the correctness and effectiveness of the proposed method in rotor and bearing fault diagnosis of induction motor. Considering that the feature extraction process is too complex in the fault diagnosis, Wen et al. [36] proposed a new convolutional neural network fault diagnosis method based on LeNet-5. Compared with the traditional p convolutional neural network, sparse filter, deep belief network, and support vector machine, the simulation results show that the proposed fault diagnosis method has significant improvements. Verbert et al. [37] analyzed the impact of uncertainty on knowledge-based fault diagnosis, compared the differences between Bayesian framework and Dempster Shafer framework in processing features and objectives. The authors investigated the effectiveness of different reasoning methods in specific applications from the application level of knowledge-based fault diagnosis methods.

1.2.3 Signal-Processing-Based Method

It is often unrealistic to establish precise mathematical models for some system processes with dynamic performance. However, considering that the system processes will generate various signals, fault diagnosis method based on signal processing has been widely studied. This method uses the signal processing methods such as wavelet transform, spectrum analysis and principal component analysis to analyze and process the measured values of the obtained signals, and extracts the characteristic values of the measurable signal's amplitude, variance, and frequency for fault diagnosis [38]. Commonly used methods based on signal processing include wavelet analysis, spectrum analysis, and information fusion. Because some systems in reality are very complex and have strong nonlinearities, it is difficult to obtain their precise mathematical models, and the fault diagnosis methods based on signal processing can effectively solve this problem. It has the advantages of good real-time performance and simple implementation. It is widely used in large-scale chemical production process which can not establish mathematical model. But at the same time, this kind of method is not effective for early potential faults. It is only effective when the fault occurs to a certain extent and affects external characteristics. In most cases, it is difficult to locate faults. Therefore, this kind of method is often combined with other fault diagnosis methods in order to improve the performance of fault diagnosis.

It is pointed out in literature [39] that signal processing is an important task in machine fault diagnosis. In view of some major problems in machine fault diagnosis, a new hybrid deep signal processing method is applied to bearing fault diagnosis, and the vibration analysis technology is combined with deep learning to form a deep learning structure embedded with time synchronous resampling mechanism. Finally, the effectiveness of the method is verified by the simulation of real data. In Kou et al. [34], a fault diagnosis method based on deep feedforward network and wavelet transform is proposed in reference to the fault of power electronics converters. The method uses wavelet transform to remove feature redundant data and compresses training samples, which greatly improves the training speed of deep feedforward networks. Experimental results show that this method can accurately locate open-circuit faults in IGBTs. In order to diagnose the operating status of wind turbines, Wang et al. [40] proposed a wind turbine gearbox fault diagnosis method based on the combination of spectrum analysis and vibration spectrum analysis. Through the vibration waveform spectrum analysis method, the severity of the system fault can be obtained and the cause of the problem can be analyzed. Simulation shows this method can realize the wind turbine gearbox fault diagnosis and ensure the reliability of the wind system.

1.3 Fault Classification

According to different classification standards, faults can be divided into different forms.

1. According to the different parts of fault, it can be divided into:

   (a) Component fault [41]: refers to the abnormality of some components or even subsystems in the controlled object, which makes the entire system unable to perform the established functions normally.

(b) Sensor fault [42]: refers to the sensor used for measurement in the control loop is in the condition of stuck, constant gain change or constant deviation, and then the measured information cannot be accurately obtained, which is specifically expressed as the difference between the measured value and actual value of the object variable.

(c) Actuator fault [43]: refers to the actuator is in the condition of stuck, constant gain change or constant deviation, and the control command cannot be executed correctly. There is difference between the input command and the actual output of the actuator.

2. According to the time characteristics, it can be divided into:

(a) Sudden change fault [44]: a fault in which the parameter value suddenly deviates greatly.

(b) Slow change fault [45]: a fault in which the parameter changes slowly with the change of time or environment.

(c) Clearance fault [46]: a fault that occurs from time to time due to aging, insufficient tolerance or poor contact.

3. According to the different forms of occurrence, it can be divided into:

(a) Additive fault [47]: refers to the unknown input acting on the system, which is zero during the normal operation of the system. Once it appears, the output of the system will change independently of the known input.

(b) Multiplicative fault [48]: refers to the change of some parameters of the system. They can cause changes in the output of the system, and these changes are also affected by the known input.

1.4   An Overview of Fault Diagnosis Process

Fault diagnosis is a means to evaluate the overall operation and fault state of the system by using all kinds of information existing in the system. In the fault diagnosis process, various inspection and test methods can be used to detect whether there is fault in the system and equipment, and then the fault location and specific fault type or value can be further determined in the fault system. The process of fault diagnosis can include three parts: fault detection, fault isolation and fault identification, and there is a progressive relationship among each part.

1.4.1   Fault Detection

Fault detection refers to the process of detecting whether a fault occurs in a system and determining the time when the fault occurs. Fault detection is usually the first step in the fault diagnosis process, and is the premise of the subsequent fault isolation and identification. If no fault is detected in the system, it indicates that the system is in normal working condition, and the system will continue to maintain a good operating state. If the system fault is detected, appropriate means can be selected to further analyze the location, type and value of the system fault. In the early fault diagnosis research, it can only be judged whether the system has a fault, that is, only the fault detection part can be realized. Later, with the development of fault diagnosis technology, the subsequent fault analysis has been gradually developed and applied.

In order to ensure the safety of wind turbine operation, Liu et al. [49] proposed a small-sample wind turbine fault detection method based on generative adversarial nets, which solves the problem of limited fault information caused by small sample fault data. Based on the actual data from wind farm, comparative experiments of wind turbine fault detection is carried out, and it is verified that the proposed method can detect the fault of wind turbine system correctly. When dealing with the inter-turn fault, which is one of the most common faults in permanent-magnet synchronous machine, wavelet transform is used to highlight the fault feature components firstly, and then the inter-turn fault is detected based on the signal analysis results of fast Fourier transform. Simulation and experimental results verify the effectiveness of this method in the incipient stage inter-turn fault [50]. The existing advanced fault detection methods are studied, classified and analyzed in literature [51], and the compatibility of each fault detection technology is studied, such as detection time, sensor requirement, program complexity, detection variables and level of protection achieved. It has important reference value for improving the possibility of photovoltaic system fault detection.

There are usually three indexes to judge the performance of fault detection in a fault diagnosis method.

1. Sensitivity: refers to the ability of a fault detection system to detect "small" fault signals. The higher the sensitivity of the detection system, the smaller the minimum fault signal it can detect.
2. Timeliness: the ability of the detection system to detect the fault as soon as possible after the fault occurs. The better the timeliness of the fault detection, the shorter the time interval between fault occurrence and fault detection.
3. False alarm rate and missing detection rate: false alarm refers to a situation that the system has no fault but has been wrongly determined to have a fault. Missing detection refers to the situation that a fault has occurred in the system but has not been detected. A reliable fault detection system should keep the false alarm rate and missing detection rate as low as possible.

1.4.2 Fault Isolation

Fault isolation is the process of designing a suitable method to locate the system fault after determining that the system has failed. Fault isolation is used to determine the working status of the subsystem or component of the system or equipment that has failed, further narrow the scope of the system fault, determine the specific fault location, and then pave the way for the specific analysis of the fault type or value.

In the fault diagnosis of nonlinear systems with sensor biased faults, Zhang et al. [52] designed a fault diagnosis architecture consists of a fault detection estimator and a bank of isolation estimators. Simulation examples show the effectiveness of the proposed sensor biased fault isolation method. Aiming at the fault diagnosis problem of modern industrial systems, Ji et al. [53] proposed an incipient fault isolation method for sensors based on the available fault direction information on the basis of augmented Mahalanobis distance analysis. This book theoretically analyzes the fault isolation conditions and compares the proposed method with conventional methods. The application in the fault diagnosis of high-speed train air brake system and continuous stirred tank reactor verifies the effectiveness of the proposed fault detection and isolation method based on the augmented Mahalanobis distance. According to the knowledge of space projection, the residual evaluation and contribution plot are unified in a framework in literature [54]. After fault detection, the process fault isolation scheme is obtained according to the optimal residuals. The simulation results in the numerical model and the Tennessee Eastman process show that the fault method can accurately locate the fault, and compared with the PCA-based fault isolation method, the proposed method has superior fault isolation performance.

Fault isolation capability refers to the ability of diagnosis system to distinguish different faults. The strength of this ability depends on the physical characteristics of the object, fault size, noise, interference, modeling error and the designed diagnosis algorithm. The stronger the separation ability is, the more accurate the fault location is.

1.4.3 Fault Identification

Fault identification is the part to determine the size, type and attribute of fault after fault detection and isolation. After the fault identification, the specific fault cause, fault type and fault size of the system can be obtained, which provides detailed fault information for the fault repair part after fault diagnosis.

In order to solve the problem that traditional contribution plots can not be applied to nonlinear process, Liu et al. [55] extended the classical fault identification method based on reconstruction to kernel independent component analysis, and developed the fault reconstruction method from unidimensional faults to multidimensional ones. The method is used to indicate the adjustment magnitude of fault sample returning to normal range as an index of fault identification, which is more direct than the existing fault identification index. Finally, the simulation results show that the method is feasible and effective in dealing with sensor faults and complex process faults. The closed-loop control system usually has the characteristics of fault propagates inside the system, and it is prone to the situations that the magnitude of the fault becomes smaller and the difference in fault characteristics is not obvious. In order to solve this problem, based on the theoretical analysis results of the influence of fault propagation on system variables, Sun et al. [56] used deep neural network to find the fault characteristics difference between different data modes, and then used a sliding window to amplify the fault noise ratio and characteristics difference, which effectively improved the fault recognition performance of the system. In literature [57], the improved weighted contribution analysis method and the method based on sensor validity index were compared in the fault diagnosis problem of a nuclear power plant. The simulation results show that the improved weighted contribution analysis method has better fault recognition performance both for single and double sensor faults. The method based on sensor validity index can not only verify the fault recognition results based on the weighted contribution analysis method, but also can reconstruct the measurement of faulty sensor according to requirements.

The accuracy of fault identification refers to the accuracy of fault size and its time-varying characteristics estimated by the diagnosis system. The higher the accuracy of fault identification, the more accurate the fault estimation of the diagnosis system, and the more conducive to fault evaluation and decision-making.

1.5   Summary of Filtering Methods

In the process of signal generation, transmission and reception, it will inevitably be affected by the interference of the external environment and the noise of the internal equipment. In order to obtain the effective estimation of the demand signal or state, it is necessary to remove noise from the mixed signals and extract useful signals. This process is called filtering. If the nature of the signal is different, the filtering method will be different.

Wiener, known as the pioneer of random process and noise signal processing, proposed the Wiener filtering theory [58]. This theory realizes the design of the filtering by analyzing the power spectrum, but it requires too much calculation of the Wiener equation, and needs infinite historical data, so it can not handle real-time data, and it is not suitable for the filtering of multi-variable, time-varying, non-stationary signals. Therefore, the above-mentioned limitations make Wiener filtering greatly restricted in practical applications. With the development of filtering theory, although the modern Wiener filtering algorithm can realize the processing of multi-dimensional and non-stationary random signals, the filtering still has the problem of non-recursive and large calculation amount, and the application scope is still limited.

In order to solve the above-mentioned problems, Kalman proposed a recursive filter estimation algorithm that uses system state equations, measurement equations and related statistical characteristics, namely the Kalman filter algorithm [59, 60]. Through this algorithm, the required signal can be estimated from the measurement data. Kalman filter provides an optimal solution to the linear Gaussian problem. In recent years, Kalman filter has been widely used in the fields of target tracking, robot control and fault diagnosis. The extended Kalman filter and the unscented Kalman filter are designed in literature [59]. The residuals from the nonlinear Kalman filter are obtained using the measurement data of some important parameters of the engine, and then the multiple model method is used to detect and diagnosis the fault of the engine components. Based on the simulated measurement data from the mathematical model of the engine, the feasibility and effectiveness of the proposed fault detection and diagnosis algorithm based on nonlinear Kalman filter method in the fault diagnosis process of the open-cycle liquid propellant rocket engine are verified by numerical simulation [61].

The particle filter [61] is based on the non-parametric Monte Carlo simulation method to realize the recursive Bayesian estimation, and describes the posterior probability distribution of the system state in the form of samples. The characteristic of particle filtering is that it needs to sample a large number of particles for estimation, which makes the algorithm have problems such as large amount of calculation and poor real-time performance. But particle filter can be widely used to deal with nonlinear and non Gaussian filtering problems, which has attracted the attention of experts and scholars in many disciplines and the improved algorithm for interdisciplinary fusion is proposed. Li et al. [61] used sequential importance re-sampling particle filtering state estimation and smoothed residual to deal with the prognostic and health management for more electric aircraft, and the simulation results verified this method can locate the faults accurately and quickly.

The above-mentioned fault diagnosis method based on filtering requires that the process interference and measurement noise of the system meet the requirements of a specific distribution law. However, in the actual systems, there will inevitably be a variety of noise interference. And in most cases, there is not enough data to summarize the random characteristics of these noises, and even some noises do not have random characteristics, so it is difficult to describe them with statistical laws. If these methods are used to solve the problem of fault diagnosis in practical system, it may lead to false alarm or missed detection. Therefore, considering the existed above problems, these methods have some limitations in practical applications.

To solve this problem, the set-membership filtering estimation method came into being, and attracted the attention of experts and scholars. The set-membership filtering estimation method describes uncertain noise as additive bounded noise with unknown probability distribution, without assuming that the noise meets the requirements of a specific distribution law. The basic principle of the set membership filtering estimation method can be described as: a feasible set which can approximately describe the real state of the system can be obtained by applying the system model, measurement data and noise boundary. The shape and complexity of the feasible set depend on the noise and the structure of system model. The set-membership filtering estimation method requires the system noise to be bounded, which makes the feasible set can be expressed as a convex polyhedron in space. Due to the complex shape of the convex polyhedron, it can be approximated by regular spatial geometry. According to the different enveloping methods, the spatial geometric shapes commonly used to describe the feasible set are: ellipsoids [62–65], interval [66–68], zonotopes [69, 70], polytopes [71, 72], orthotopes [73, 74] and so on.

Compared with the traditional filtering method, the advantages of the set-membership filtering estimation method are as follows [75–77]:

1. The traditional filtering method assumes that the system noise is random noise, and it is necessary to know the probability distribution of the noise, which is often difficult to obtain in the actual system. The set-membership filtering method only requires the noise to be bounded, which is closer to the actual situation. Therefore, compared with the traditional filtering method, the set- membership filtering method is more universal and practical.

2. Since the modeling uncertainty of the system can also be regarded as a kind of bounded noise, it is enough to obtain the equivalent bounded noise characteristics of the uncertain system when dealing with the parameter uncertain system. Therefore, the set-membership filtering method can conveniently handle the situation where the system has unmodeled dynamics.

3. Set-membership filtering only requires the noise to be bounded, and the upper and lower bounds are known. There is no requirement for the specific distribution of noise. Therefore, compared with other traditional filtering methods, this method has stronger robust performance.

Due to the strong robustness and good practicability of set-membership filtering estimation method, it has gradually become a research hot-spot in recent years, and has a wide range of applications in fault diagnosis, system modeling and pattern recognition.

The shape of ellipsoid is regular and its calculation is simple, so it is a kind of algorithm that has attracted much attention from scholars. Schweppe is the scholar who studies the ellipsoid set-membership estimation algorithm in the early stage. In 1968, Schweppe adopted an ellipsoid approximation to describe the feasible set of states, and gave an ellipsoid set-membership filtering algorithm [78]. In 1982, Fogel and Huang introduced the ellipsoid into the approximate description of the feasible set of parameters, and proposed the ellipsoid set-membership filtering estimation algorithm [79]. Subsequently, the ellipsoid based set-membership filtering estimation method has been developed rapidly, and fruitful research results have been achieved, and a series of practical problems in fault diagnosis have been solved.

Xia et al. [80] studied a distributed network set-membership filtering problem with ellipsoidal state estimation, which solved the state estimation problem of discrete time-varying systems with unknown but bounded processes and measurement noise. The experiment of the single-phase grid-connected power generation system platform verifies the feasibility and effectiveness of the method in practical application. Zhou et al. [81] regarded the model uncertainty and parameter variation as bounded errors, and proposed a set membership identification method based on ellipsoidal boundaries, which effectively includes fault parameters under real parameter values. This method has improved algorithm robustness and can determine the difference between actual fault and model uncertainty. The simulation results of a mobile robot with multiple sliding fault scenarios verify the correctness of the proposed method in fault detection and fault isolation. Huang et al. [82] proposed a fault diagnosis method based on the optimal ellipsoid for the fault detection and isolation of the robotic assembly of electrical connectors. If the feasible parameters of the fault-free switching linear model calculated based on the optimal ellipsoid are inconsistent with any possible sub-model, a system fault is detected. Then the fault isolation is realized by checking the consistency of the data sequence and the possible fault model one by one. The simulation experiment based on the robotic assembly experiments of mating electrical connectors verifies the effectiveness of the proposed algorithm.

Interval analysis uses intervals to realize the storage operation of data, and the operation result is guaranteed to contain all possible true values, that is, the result is accurate and reliable. In addition, people can easily express certain uncertainty calculation parameters as intervals and directly include them in the interval algorithm, which is also of great significance in practical applications.

Jaulin and Walter are the first scholars to apply interval analysis to set-membership filtering estimation. In 1993, Jaulin and Walter proposed a set inverse via interval analysis (SIVIA) algorithm that can handle the set-membership estimation problem of nonlinear systems [83]. Since 2000, set-membership filtering algorithms based on interval analysis have also been proposed one after another. Due to the need for dichotomy, SIVIA is not suitable for processing high-dimensional data. In order to solve this problem, the improved algorithm which can overcome the dimensionality disaster is proposed in the literature [84, 85].

The set-membership filtering algorithm based on interval estimation can deal with the set-membership estimation problem of nonlinear systems, and the effect is very significant. In addition, compared with most other set-membership methods, this method can guarantee the global optimization. Therefore, the set-membership filtering method based on interval estimation is also widely used in fault diagnosis. Aiming at the fault diagnosis problem of uncertain discrete linear systems, Meslem et al. [66] designed a set-membership state estimator based on interval calculation to deal with system outliers, which effectively guaranteed the convergence of the estimated state enclosures width and the algorithm robustness against outliers in data. Finally, two numerical simulations show the effectiveness of the proposed set-membership state estimator based on interval computation. In order to solve the problem of wind turbine fault diagnosis, Sanchez et al. [67] proposed a model-based fault diagnosis method by using intervals to describe the uncertainty of model parameters and the unknown but bounded noise as well as combining analytical redundancy relations. The fault detection of this method is realized by checking whether the measured value falls within the estimated output interval, and the fault isolation is realized based on considering a set of ARRs obtained from the structural analysis of the wind turbine model and a fault signature matrix that considers the relation of ARRs and faults. Finally, the proposed method has been simulated by a 5-MW wind turbine. In literature [68], the uncertain interference and noise are included in the bounded set, and the characteristics of minimum detectable faults are described by using the residual sensitivity and set invariance theory. The proposed interval observer based fault detection method is verified by a quadruple-tank system.

A zonotope is the affine transformation of a unit hypercube. Boxes and parallelotopes can be regarded as special cases of zonotopes. In fact, the three are all affine transformations of the unit hypercube, but the form of the affine transformation matrix is different. If the affine transformation matrix is a diagonal matrix, a full-rank square matrix, and a row full-rank matrix, the result is a box, a parallelotope, and a zonotope, respectively. It can be seen that the zonotope based set-membership estimation algorithm will be more powerful than the box and parallelotope based set-membership estimation algorithms. In addition, since the unit hypercube can also be regarded as a unit interval vector, the concepts and methods in interval analysis will be used when constructing the zonotope based set-membership estimation algorithm.

In 1998, Kühn pointed out that the warping effect can be controlled effectively by using zonotopes in literature [86]. After that, zonotope based set-membership filtering algorithms suitable for different systems have been constructed continuously, and the fault diagnosis problems of different systems have been effectively dealt with. Pourasghar et al. [68] used zonotopes to contain uncertain state estimation sets, analyzed and compared the performance differences in state estimation and fault detection using interval observers and set-membership estimation methods in uncertain linear systems, which is verified in the simulation of a dual tank system. In order to improve the sensitivity of the system faults relative to interference, Masoud et al. [87] proposed a new type of fault detection observer based on zonotopic Kalman filter. Simulation experiments and analysis show that this algorithm can improve the sensitivity of system to faults and enhance the robustness of the system to interference. Considering the influence of modeling uncertainty, the zonotope is used to contain the uncertain parameter set in Blesa et al. [88], and two set-membership identification methods are proposed: interval predictor method and bounded error method. At the same time, two robust fault detection tests which are generated by the two identification methods are discussed and the basic assumptions and application conditions are given. Finally, a four-tank system is taken as an example to illustrate the applicability and characteristics of the two proposed identification and fault detection methods.

Comparing the several algorithms introduced above, we can find their respective characteristics. The set-membership estimation algorithms based on ellipsoid, box, approximate or exact polytope, hyperplane, and zonotope generally use a single geometric body to approximately or accurately describe the feasible set. Among them, the ellipsoid and parallelotope based set-membership estimation algorithms are generally recursive algorithms. The calculation process is relatively simple, but the accuracy is relatively low. The zonotope based set-membership estimation algorithm is also a recursive algorithm, the calculation process is slightly more complicated, but the accuracy is improved. The approximate or exact polytope based set-membership estimation algorithm is also a recursive algorithm with the highest accuracy, but the calculation process is more complicated and requires a larger computer storage capacity. The box based set-membership estimation algorithm is usually a batch algorithm with moderate accuracy, but it also has the problem of large amount of calculation. Compared with the above algorithms, the set-membership estimation algorithm based on interval analysis has a completely different idea in approximate description of feasible set. The result of the algorithm is a union of multiple geometry instead of a single geometry, and this union can be approximated to an approximate feasible set with any accuracy. However, due to the shortcomings of interval analysis, the convergence speed of the algorithm may not be good.

The fault detection process based on set-membership estimation is generally completed by detecting whether the approximate feasible set is empty. The fault detection strategy can be expressed as: if the approximate feasible set is detected as an empty set, then it is determined that the system is faulty, otherwise, it is determined that the system is not faulty. This strategy is based on the following facts: when the system is fault-free, the approximate feasible set must not be empty, and if the approximate feasible set is detected as an empty set, the system must fail. Therefore, the false detection rate of the fault detection method based on the collective estimation must be zero, that is, once a fault alarm is given, there must be a fault. In order to detect the fault of the system with bounded parameter variation between samples, the polytopes are used in literature [89] to contain the parameter uncertainty set, and the consistency checks indicates whether the system fails. Finally, an example of the four-tank system is used to show the fault detection process and results of the algorithm. In view of the problem of high missed detection rate in conventional set-membership estimation methods, Wan et al. [90] proposed a probabilistic set-membership parity relation method based on the probabilistic information on the parametric uncertainties, which results in a reduced missed detection rate by admitting an acceptable false alarm rate. In the test of fault consistency, a non-convex confidence set of residuals is determined. The effectiveness of the proposed method is verified in the fault detection case of continuous stirred tank reactor.

In the research of fault isolation and fault identification process based on set-membership estimation. In Fernandez-Canti et al. [91], the modeling error is regarded as unknown but bounded, and the measurement noise is assumed to be bounded and follow the statistical distribution within the bound. A fault detection and isolation method based on mixed Bayesian/Set-membership method is proposed to deal with the problem of wind turbine fault. Based on the Bayesian theory and its fault isolation framework, a new fault isolation method is developed. And the fault isolation process is realized by matching the test results of fault detection. For a class of multi-rate time-varying systems with sensor degradations, unknown but bounded disturbance and faults, Zhang et al. [92] studied the problem of fault diagnosis, and proposed a new event-triggered communication mechanism. In this method, the fault isolation estimator is implemented based on set-membership estimation and recursive matrix inequalities. After fault detection, the proposed estimation-error-function matching approach is used to realized the fault isolation. Finally, the effectiveness of the proposed method is verified by numerical simulation in a three tank system. The model uncertainties and parameter variations are regarded as bounded errors by Zhou et al. [81], a set-membership identification algorithm with strictly propagating bounded uncertainty is proposed by using ellipsoid to contain the feasible set of system fault parameters. This method can detect and isolate the fault arisen from abrupt parameter variations, and can effectively distinguish the actual fault from the model uncertainties. Based on the resuming and better formalizing the two complementary definitions of set-membership identifiability and mu-set-membership identifiability, literature [93] proposed a fault detection and identification method. This method is based on differential algebra and uses the relationship between the observations, inputs, and unknown parameters of system to estimate the uncertain parameters of the model to realize the fault identification process.

1.6 Motivation and Objective

Based on the analyses of the above-mentioned literatures, filtering theory and methods have been widely developed in recent decades, and have been highly valued by scholars. Filtering theory is rich in content, and is expanding to many aspects. There are many innovative research results in the application of image processing and recognition, fault diagnosis, predictive control and so on. As one of the important application fields of filtering theory, fault diagnosis can be carried out by designing suitable filters, and it has been successfully applied to the fault diagnosis research of electric power, chemical industry, machinery and other industries. In general, although the results of fault diagnosis based on filtering are relatively rich, the theoretical research is far from perfect, and a lot of research work still needs to be done. At present, although a lot of research results have been achieved in the research of filtering-based fault diagnosis method, the traditional filtering-based fault diagnosis methods mostly require that the system noise meet the specific distribution law, such as Kalman filter and particle filter, which makes it suffer great limitations in practical application. In order to solve this problem, set-membership filtering estimation method arose at the historic moment, and effectively broadened the practical application field of fault diagnosis method based on filtering.

For the fault diagnosis method based on set-membership filtering, which can handle the fault diagnosis of systems with unknown but bounded noise, there is still room for further improvement and expansion in the theoretical research. For example, for the system with unknown noise, there are few research results on abrupt fault diagnosis with low computational load and high diagnostic efficiency. At present, model matching is usually used in the diagnosis of abrupt faults, which can also obtain a good fault diagnosis results. However, for the system with more fault types, there are some limitations. Once the matching sequence is inconsistent with the actual sequence of faults, the result can only be obtained when the last type of failure is matched. This situation will reduce the operating efficiency of the algorithm and limit the diagnosis speed of the abrupt faults. For another example, the existing fault diagnosis tends to focus on the type of sudden change, and there is less research on slow change faults whose impact cannot be ignored while the magnitude of the fault continues to increase. However, the slow change faults are the roots cause that affects the service life of industrial equipment and brings uncontrollable production safety. Therefore, how to realize the effective diagnosis of slowly changing fault is an important problem to be solved in the current field of fault diagnosis.

In summary, we must continue to research and explore in this field to improve the theoretical system of fault diagnosis method based on set-membership filtering. On the basis of the research results of existing scholars, it is of great practical significance to continue to carry out the study and research to supplement and improve the theoretical system of fault diagnosis based on set-membership filtering and apply it to the practical engineering field.

1.7 Outlines

Although the research work on set membership estimation theory, methods and their applications has achieved some results, there is a lot of work to be done. Based on analyzing the existing results, further research work is carried out in this dissertation. The research work here focuses on existing methods improvement, new methods presentation, and applications of set membership estimation theory and methods to fault diagnosis.

The following contents and arrangements of this book are as follows. Chapter 2, studies and analyzes the traditional fault diagnosis method based on filtering, and the Kalman filter algorithm is taken as an example to simulate and verify in the fault diagnosis of power converter. In Chap. 3, the fault detection method of set-membership filtering based on ellipsoid is analyzed and studied. Aiming at the problems of repeated calculation and low data utilization rate of the parameter estimation algorithm of weight ellipsoid, the parameter estimation algorithm of limited data window of weight ellipsoid is proposed. In Chap. 4, based on the polyhedral set-membership algorithm, a fault detection algorithm based on polyhedral set-membership filtering is proposed. In Chap. 5, based on the knowledge of interval operation and the idea of SIVIA, a fault observer based on vector set inversion interval filtering is designed for the fault detection of DC motor. In Chap. 6, based on zonotopes and orthotopes, combined with linear programming theory, the fault diagnosis method of set-membership filtering based on polyhedron is studied. At the same time, considering the directional expansion theory and the stratification idea, the targeted research was carried out respectively. In Chap. 7, two fault diagnosis methods based on composite set-membership filtering are proposed. Finally, in Chap. 8, the research contents of this book are summarized, and the future development direction of fault diagnosis method based on set-membership filtering is prospected.

References

1. Y.Z. Li, Z.X. Ni, T.Y. Zhao, M.H. Yu, Y. Liu, L. Wu, Y. Zhao, Coordinated scheduling for improving uncertain wind power adsorption in electric vehicles-wind integrated power systems by multiobjective optimization approach. IEEE Trans. Ind. Appl. **56**(3), 2238–2250 (2020)
2. Z.Q. Geng, Z. Wang, H.X. Hu, Y.M. Han, X.Y. Lin, Y.H. Zhong, A fault detection method based on horizontal visibility graph-integrated complex networks: application to complex chemical processes. Can. J. Chem. Eng. **97**(5), 1129–1138 (2019)
3. V.G. Cannas, J. Gosling, M. Pero, T. Rossi, Engineering and production decoupling configurations: an empirical study in the machinery industry. Int. J. Prod. Econ. **216**, 173–189 (2019)

4. C. Edwards, S. Simani, Fault diagnosis and fault-tolerant control in aerospace systems. Int. J. Robust Nonlinear Control **29**(16), 5291–5292 (2019)
5. S.J. Ma, W Cai, W.K. Liu, Z.W. Shang, G. Liu, A lighted deep convolutional neural network based fault diagnosis of rotating machinery. Sensors **19**(10) (2019)
6. D. Gielen, F. Boshell, D. Saygin, M.D. Bazilian, N. Wagner, R. Gorini, The role of renewable energy in the global energy transformation. Energ. Strat. Rev. **24**, 38–50 (2019)
7. M. Balat, A review of modern wind turbine technology. Energy Sour. Part A-Recov. Util. Envion. Effects **31**(17), 1561–1572 (2009)
8. G.Q. Jiang, H.B. He, J. Yan, P. Xie, Multiscale convolutional neural networks for fault diagnosis of wind turbine gearbox. IEEE Trans. Ind. Electron. **66**(4), 3196–3207 (2019)
9. J.H. Lei, C. Liu, D.X. Jiang, Fault diagnosis of wind turbine based on long short-term memory networks. Renew. Energy **133**, 422–432 (2019)
10. R. Umaz, A two-stage power converter architecture with maximum power extraction for low-power energy sources. Turk. J. Electr. Eng. Comput. Sci. **27**(6), 4744–4755 (2019)
11. X.X. Zheng, P. Peng, Fault diagnosis of wind power converters based on compressed sensing theory and weight constrained adaboost-svm. J. Power Electron. **19**(2), 443–453 (2019)
12. A. Ismail, L. Saidi, M. Sayadi, Wind turbine power converter fault diagnosis using dc-link voltage time-frequency analysis. Wind Eng. **43**(4), 329–343 (2019)
13. R.V. Beard, Failure accomodation in linear systems through self-reorganization. Massachusetts Institute of Technology (1971)
14. A.S. Willsky, A survey of design methods for failure detection in dynamic systems. Automatica **12**(6), 601–611 (1975)
15. D.M. Himmelblau, *Fault Detection and Diagnosis in Chemical and Petrochemical Process* (Elsevier Press, Amsterdam, 1978)
16. B. Jiang, P. Shi, Y. Wu, Incipient fault diagnosis for t-s fuzzy systems with application to high-speed railway traction devices
17. Y.Y. Dai, Y. Qiu, A stacked auto-encoder based fault diagnosis model for chemical process. Comput. Aided Chem. Eng. **46**, 1303–1308 (2019)
18. J. Li, H.L. Jiang, G.H. Liu, Model based fault diagnosis for drillstring washout using iterated unscented kalman filter. J. Petrol. Sci. Eng. **180**, 246–256 (2019)
19. H. Xia, A. Ayodeji, Y.K. Liu, Knowledge base operator support system for nuclear power plant fault diagnosis. Prog. Nucl. Enegr. **105**, 42–50 (2018)
20. B. De Schutter, K. Verbert, R. Babuska, Bayesian and dempster-shafer reasoning for knowledge-based fault diagnosis-a comparative study. Eng. Appl. Artif. Intel. **60**, 136–150 (2017)
21. D.H. Zhou, Y. Liu, Z.D. Wang, State estimation and fault reconstruction with integral measurements under partially decoupled disturbances. IET Control Theory Appl. **12**(10), 1520–1526 (2018)
22. H. Al Samrout, C. Delpha, D. Diallo, Multiple incipient fault diagnosis in three-phase electrical systems using multivariate statistical signal processing. Eng. Appl. Artif. Intel. **73**, 68–79 (2018)
23. Y. Qu, M. He, D. He, A new signal processing and feature extraction approach for bearing fault diagnosis using ae sensors. J. Fail. Anal. Prev. **16**(5), 1–7 (2016)
24. X. Zhang, F. Ding, E.F. Yang, State estimation for bilinear systems through minimizing the covariance matrix of the state estimation errors. Int. J. Adapt. Control Signal Process. **33**(7), 1157–1173 (2019)
25. X. Zhang, F. Ding, L. Xu, Recursive parameter estimation methods and convergence analysis for a special class of nonlinear systems. Int. J. Robust Nonlinear Control **30**(4) (2019)
26. S. Cho, J. Jin, Optimal fault classification using fisher discriminant analysis in the parity space for applications to npps. IEEE Trans. Nucl. Sci. **65**(3), 856–865 (2018)
27. K.O. Mtepele, D.U. Campos-Delgado, A.A. Valdez-Fernández, et al, Model-based strategy for open-circuit faults diagnosis in n-level CHB multilevel converters. IET Power Electron. **12**(4), 648–655 (2018)
28. A.J. Yan, Y.J. Wang, D.H. Wang, Fault diagnosis method using learning case-based reasoning for Tennessee-Eastman process. Control Theory Appl. **34**(9), 1179–1184 (2017)

29. Q. Chen, J.C. Wang, A. Qadeer et al., Model-based fault diagnosis of automotive electric power steering system. Auto. Eng. **41**(7), 839–850 (2019)
30. S. Huang, K.K. Tan, T.H. Lee, Fault siagnosis and fault-tolerant control in linear drives using the kalman filter. IEEE Trans. Ind. Electron. **59**(11), 4285–4292 (2012)
31. Z.T. Liu, H.W. He, Sensor fault detection and isolation for a lithium-ion battery pack in electric vehicles using adaptive extended kalman filter. Appl. Energy **185**, 2033–2044 (2017)
32. X.F. Lv, D.Y. Zhou, L. Ma, Y.C. Tang, Dependency model-based multiple fault diagnosis using knowledge of test result and fault prior probability. Appl. Sci.-Basel **9**(2) (2019)
33. Z. Gao, C. Cecati, S.X. Ding, A survey of fault diagnosis and fault-tolerant techniques-part ii: fault diagnosis with knowledge-based and hybrid/active approaches. IEEE Trans. Industr. Electron. **62**(6), 3768–3774 (2015)
34. L. Kou, C. Liu, G.M. Cai, J.N. Zhou, Q.D. Yuan, S.M. Pang, Fault diagnosis for open-circuit faults in npc inverter based on knowledge-driven and data-driven approaches. IET Power Electron. **13**(6), 1236–1245 (2020)
35. J.H. Lee, J.H. Pack, I.S. Lee, Fault diagnosis of induction motor using convolutional neural network. Appl. Sci.-Basel **9**(15) (2019)
36. L. Wen, X.Y. Li, L. Gao, Y.Y. Zhang, A new convolutional neural network-based data-driven fault diagnosis method. IEEE Trans. Ind. Electron. **65**(7), 5990–5998 (2018)
37. K. Verbert, R. Babuska, B. De Schutter, Bayesian and dempster-shafer reasoning for knowledge-based fault diagnosis-a comparative study. Eng. Appl. Artif. Intell. **60**, 136–150 (2017)
38. S. Choi, B. Akin, M.M. Rahimian, H.A. Toliyat, Implementation of a fault-diagnosis algorithm for induction machines based on advanced digital-signal-processing techniques. IEEE Trans. Ind. Electron. **58**(3), 937–948 (2011)
39. M. He, D. He, A new hybrid deep signal processing approach for bearing fault diagnosis using vibration signals. Neurocomputing **396**, 542–555 (2020)
40. S.B. Wang, B. Zhao, Y.S. Luo, Wind turbine gearbox fault diagnosis based on the vibration spectrum analysis. J. Comput. Methods Sci. Eng. **19**(1), 137–151 (2019)
41. M. Saimurugan, K.I. Ramachandran, V. Sugumaran, N.R. Sakthivel, Multi component fault diagnosis of rotational mechanical system based on decision tree and support vector machine. Expert Syst. Appl. **38**(4), 3819–3826 (2011)
42. B. Gou, X. Yan, X. Yang, G. Wilson, S.Y. Liu, An intelligent time-adaptive data-driven method for sensor fault diagnosis in induction motor drive system. IEEE Trans. Ind. Electron. **66**(12), 9817–9827 (2019)
43. N. Pizzi, E. Kofman, J.A. De Dona, M.M. Seron, Actuator fault tolerant control based on probabilistic ultimate bounds - sciencedirect. ISA Trans. **84**, 20–30 (2019)
44. H. Yoshida, S. Kumar, Arx and afmm model-based on-line real-time data base diagnosis of sudden fault in ahu of vav system. Energy Convers. Manage. **40**(11), 1191–1206 (1999)
45. Y. Song, S. Yang, C. Cheng, P. Xie, A novel fault detection method for running gear systems based on dynamic inner slow feature analysis. IEEE Access **8**, 211371–211379 (2020)
46. X. Bi, S. Cao, D. Zhang, Diesel engine valve clearance fault diagnosis based on improved variational mode decomposition and bispectrum. Energies **12**(4) (2019)
47. J. Stoustrup, H. Niemann, Active fault diagnosis by controller modification. Int. J. Syst. Sci. **41**(8), 925–936 (2010)
48. H. Hao, K. Zhang, S.X. Ding, Z. Chen, Y. Lei, A data-driven multiplicative fault diagnosis approach for automation processes. ISA Trans. **53**(5), 1436–1445 (2014)
49. J.H. Liu, F.M. Qu, X.W. Hong, H.G. Zhang, A small-sample wind turbine fault detection method with synthetic fault data using generative adversarial nets. IEEE Trans. Ind. Inf. **15**(7) (2019)
50. J. Fang, Y.N. Sun, Y.B. Wang, B.L. Wei, L. Hang, Improved zsvc-based fault detection technique for incipient stage inter-turn fault in pmsm. IET Electr. Power Appl. **13**(12), 2015–2026 (2019)
51. D.S. Pillai, F. Blaabjerg, N. Rajasekar, A comparative evaluation of advanced fault detection approaches for pv systems. IEEE J. Photovolt. **9**(2), 513–527 (2019)

52. X.D. Zhang, T. Parisini, M.M. Polycarpou, Sensor bias fault isolation in a class of nonlinear systems. IEEE Trans. Autom. Control **50**(3), 370–376 (2005)
53. H.Q. Ji, K.K. Huang, D.H. Zhou, Incipient sensor fault isolation based on augmented mahalanobis distance. Control. Eng. Pract. **86**, 144–154 (2019)
54. J. Wang, W.S. Ge, J.L. Zhou, H.Y. Wu, Q.B. Jin, Fault isolation based on residual evaluation and contribution analysis. J. Frankl. Inst.-Eng. Appl. Math. **354**(6), 2591–2612 (2017)
55. Y. Liu, F.L. Wang, Y.Q. Chang, Reconstruction in integrating fault spaces for fault identification with kernel independent component analysis. Chem. Eng. Res. Des. **91**(6), 1071–1084 (2013)
56. B.W. Sun, J.Q. Wang, Z.M. He, H.Y. Zhou, F.S. Gu, Fault identification for a closed-loop control system based on an improved deep neural network. Sensors **19**(9) (2019)
57. W. Li, M.J. Peng, Q.Z. Wang, Fault identification in pca method during sensor condition monitoring in a nuclear power plant. Ann. Nucl. Energy **121**, 135–145 (2018)
58. M. Yang, C. Jin, G. Dong, Weak fault feature extraction of rolling bearing based on cyclic wiener filter and envelope spectrum. Mech. Syst. Signal Process. **25**(5), 1773–1785 (2011)
59. J. Cha, S. Ko, S.Y. Park, E. Jeong, Fault detection and diagnosis algorithms for transient state of an open-cycle liquid rocket engine using nonlinear kalman filter methods. Acta Astronaut. **163**, 147–156 (2019)
60. F. Joyce, S. Dmitry, Wavefront reconstruction with defocus and transverse shift estimation using Kalman filtering. Opt. Lasers Eng. **11**(1), 122–129 (2018)
61. H. Li, Y. Hui, J. Qu, H. Sun, Fault diagnosis using particle filter for mea typical components. J. Eng. **13**, 603–606 (2018)
62. Z.G. Wang, X.J. Shen, Y.M. Zhu, A tighter set-membership filter for some nonlinear dynamic system. IEEE Access **6**, 25351–25362 (2017)
63. F.Q. You, H.L. Zhang, F.L. Wang, A new set-membership estimation method based on zonotopes and ellipsoids. Trans. Inst. Meas. Control **40**(7), 2091–2099 (2018)
64. T. Alamo, J.M. Bravo, E.F. Camacho, Guaranteed state estimatin by zonotope. Automatic **41**, 1035–1043 (2005)
65. Q. Shen, J.Y. Liu, X.G. Zhou, Low-complexity ISS state estimation approach with bounded disturbances. Int. J. Adapt. Control Signal Process. **32**(10), 1473–1488 (2018)
66. N. Meslem, A. Hably, Robust set-membership state estimator against outliers in data. IET Control Theory Appl. **14**(13), 1752–1761 (2020)
67. H. Sanchez, T. Escobet, V. Puig, P.F. Odgaard, Fault diagnosis of an advanced wind turbine benchmark using interval-based arrs and observers. IEEE Trans. Ind. Electron. **62**(6), 3783–3793 (2015)
68. M. Pourasghar, V. Puig, C.C. Ocampo-Martinez, Interval observer versus set-membership approaches for fault detection in uncertain systems using zonotopes. Int. J. Robust Nonlinear Control **29**(10), 2819–2843 (2019)
69. H. Wang, V.K. Ilya, J. Sun, Zonotope-based recursive estimation of the feasible solution set for linear static systems with additive and multiplicative uncertainties. Automatica **95**, 236–245 (2018)
70. Y. Wang, V. Puig, G. Cembrano, Set-membership approach and Kalman observer based on zonotopes for discrete-time descriptor systems. Automatica **93**, 35–43 (2018)
71. D. Rotondo, F. Nejjari, V. Puig, J. Blesa, Model reference ftc for lpv systems using virtual actuators and set-membership fault estimation. Int. J. Robust Nonlinear Control **25**(5), 735–760 (2015)
72. J. Hu, B. Ding, An efficient offline implementation for output feedback min-max mpc: offline min-max mpc. Int. J. Robust Nonlinear Control **29**(2), 492–506 (2018)
73. V. Reppa, A. Tzes, Fault detection and diagnosis based on parameter set estimation. IET Control Theory Appl. **5**(1), 69–83 (2011)
74. M. Casini, A. Garulli, A. Vicino, A constraint selection technique for set membership estimation of time-varying parameter. In: Proceedings of 53rd IEEE Conference on Decision and Control, vol. 32, pp. 1029–1034 (2014)
75. N. Xia, F.W. Yang, Q.L. Han, Distributed event-triggered networked set-membership filtering with partial information transmission. IET Control Theory Appl. **11**(2), 155–163 (2017)

76. A. Abur, Y. Lin, Robust state estimation against measurement and network parameter error. IEEE Trans. Power Syst. **33**(5), 4751–4759 (2018)
77. A. Garulli, M. Casini, A. Vicino, A linear programming approach to online set membership parameter estimation for linear regression models. Int. J. Adapt. Control Signal Process. **31**(3), 360–378 (2017)
78. F.C. Schweppe, Recursive state estimation: Unknown but bounded errors and system inputs. **13**(1), 22–28 (1968)
79. E. Fogel, Y.F. Huang, On the value of information in system identification-bounded noise case. Automatica **18**(2), 229–238 (1982)
80. N. Xia, F.W. Yang, Q.L. Han, Distributed networked set-membership filtering with ellipsoidal state estimations. Inf. Sci. **432**, 52–62 (2018)
81. B. Zhou, K. Qian, X.D. Ma, X.Z. Dai, Ellipsoidal bounding set-membership identification approach for robust fault diagnosis with application to mobile robots. J. Syst. Eng. Electron. **28**(5), 986–995 (2017)
82. J. Huang, Y. Wang, T. Fukuda, Set-membership-based fault detection and isolation for robotic assembly of electrical connectors. IEEE Trans. Autom. Sci. Eng. **15**(1), 160–171 (2018)
83. L. Jaulin, E. Walter, Set inversion via interval analysis for nonlinear bounded-error estimation. Automatica **29**(4), 1053–1064 (1993)
84. L. Jaulin, Interval constraint propagation with application to bounded-error estimation. Automatica **36**(10), 1547–1552 (2000)
85. L. Jaulin, M. Kieffer, I. Braems, Guaranteed non-linear estimation using constraint propagation on sets. Int. J. Control **74**(18), 1772–1782 (2001)
86. W. Kühn, Rigorously computed orbits of dynamical systems without the wrapping effect. Computing **61**(1), 47–67 (1998)
87. M. Pourasghar, C. Combastel, V. Puig, C. Ocampo-Martinez, Fd-zkf: A zonotopic kalman filter optimizing fault detection rather than state estimation. J. Process Control **73**, 89–102 (2019)
88. J. Blesa, V. Puig, J. Saludes, Identification for passive robust fault detection using zonotope-based set-membership approaches. Int. J. Adaptive Control Signal Proc. **25**(9), 788–812 (2011)
89. J. Blesa, V. Puig, J. Saludes, Robust fault detection using polytope-based set-membership consistency test. IET Control Theory Appl. **6**(12), 1767–1777 (2012)
90. Y.M. Wan, V. Puig, C. Ocampo-Martinez, Y. Wang, E. Harinath, R.D. Braatz, Fault detection for uncertain lpv systems using probabilistic set-membership parity relation. J. Process Control **87**, 27–36 (2020)
91. R.M. Fernandez-Canti, J. Blesa, S. Tornil-Sin, V. Puig, Fault detection and isolation for a wind turbine benchmark using a mixed bayesian/set-membership approach. Annu. Rev. Control. **40**, 59–69 (2015)
92. Y. Zhang, Z.D. Wang, L.E. Ma, F.E. Alsaadi, Annulus-event-based fault detection, isolation and estimation for multirate time-varying systems: applications to a three-tank system. J. Process Control **75**, 48–58 (2019)
93. C. Jauberthie, N. Verdiere, L. Trave-Massuyes, Fault detection and identification relying on set-membership identifiability. Annu. Rev. Control. **37**(1), 129–136 (2013)

Chapter 2
Design of State Space Based Fault Diagnosis Filter

2.1 Preliminaries and Problem Formulation

As one of the basic topologies of power converters, Buck converter is used for step-down in DC-DC converter field. The topology diagram of Buck converter is shown in Fig. 2.1, where i_L, u_o, L, C, and R are the inductor current, output voltage, inductor, electrolytic capacitor, and load resistor, respectively.

This chapter adopts the equivalent transformation method of non-ideal Buck converter in [1] and simplifies the circuit. The equivalent transformation method is obtained by equating the switch MOSFET to the ideal switch S_1, and the diode D to the ideal switch S_2. At the same time, the inductor is regarded as an ideal component, and the electrolytic capacitor is equivalent to the series connection of the capacitor C and the equivalent series resistor R_c. The equivalent circuit of the non-ideal Buck converter after equivalence is shown in Fig. 2.2.

When the inductor current is in the continuous conduction model (CCM for short), there are two situations in Buck converter, that is, S_1 is closed and S_2 is open, and S_1 is open and S_2 is closed [2]. Combining the states of the two cases, and using S to represent the switching state of the MOSFET, the hybrid system model of Buck converter in CCM can be obtained as

$$\begin{bmatrix} \dot{i}_L \\ \dot{u}_o \end{bmatrix} = \begin{bmatrix} 0 & -\frac{1}{L} \\ \frac{R}{C(R+R_c)} & -\frac{L+RR_cC}{CL(R+R_c)} \end{bmatrix} \begin{bmatrix} i_L \\ u_o \end{bmatrix} + S \begin{bmatrix} \frac{E}{L} \\ \frac{RR_cE}{L(R+R_c)} \end{bmatrix}. \tag{2.1}$$

Equation (2.2) can be obtained after discretizing Eq. (2.1), and T is the sampling period.

$$\begin{bmatrix} i_L(t) \\ u_o(t) \end{bmatrix} = \begin{bmatrix} 1 & -\frac{T}{L} \\ \frac{RT}{C(R+R_c)} & 1 - \frac{(L+RR_cC)T}{CL(R+R_c)} \end{bmatrix} \begin{bmatrix} i_L(t-1) \\ u_o(t-1) \end{bmatrix}$$
$$+ S(t-1) \begin{bmatrix} \frac{ET}{L} \\ \frac{RR_cET}{L(R+R_c)} \end{bmatrix}. \tag{2.2}$$

© The Author(s), under exclusive license to Springer Nature Singapore Pte Ltd. 2022
Z. Wang et al., *Advances in Fault Detection and Diagnosis Using Filtering Analysis*,
https://doi.org/10.1007/978-981-16-5959-1_2

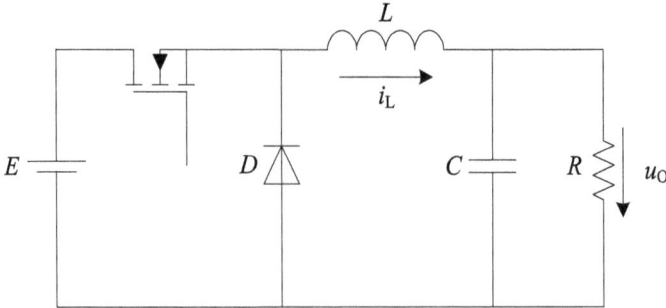

Fig. 2.1 Topology diagram of Buck converter

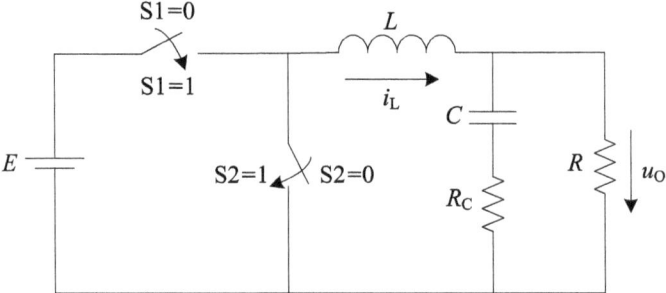

Fig. 2.2 Equivalent circuit diagram of non-ideal Buck converter

With the continuous operation of Buck converter, the components in the circuit will be aging, which will affect the normal operation of the circuit. The faults of Buck converter include inductor fault, electrolytic capacitor fault, power switch fault, etc. The electrolytic capacitor fault is the main cause of Buck converter fault among them. This chapter mainly focuses on the fault diagnosis of Buck converter caused by the degradation of electrolytic capacitor, and assumes that only one fault occurs at the same time.

2.2 Fault Diagnosis Based on Inverse Kalman Filter

Based on the hybrid system discrete model, the following matrices are defined:

$$H(t) = \left[i_L(t-1)I_2, u_o(t-1)I_2, S(t-1)I_2 \right], \tag{2.3}$$

$$\begin{aligned} X &= \left[x_{11}\ x_{12}\ x_{13}\ x_{14}\ x_{15}\ x_{16} \right]^{\mathrm{T}} \\ &= \left[1\ \ \frac{RT}{C(R+R_c)}\ \ -\frac{T}{L}\ \ 1 - \frac{(L+RR_cC)T}{LC(R+R_c)}\ \ \frac{ET}{L}\ \ \frac{RR_cET}{L(R+R_c)} \right]^{\mathrm{T}}, \end{aligned} \tag{2.4}$$

$$Y(t) = \begin{bmatrix} i_L(t) \\ u_o(t) \end{bmatrix},$$ (2.5)

then

$$Y(t) = H(t)X.$$ (2.6)

where X is the parameter matrix of Buck converter model and $H(t)$ is the state matrix of Buck converter. The purpose of parameter identification of Buck converter is to get the value of unknown parameter matrix X by using inverse Kalman filter (IKF for short) algorithm based on the known matrix $H(t)$.

Firstly, the parameter identification of Buck converter in a fault-free state is considered. For this input-output system, the unknown parameters of the system are regarded as unknown states, and then the hybrid system model expression of Buck converter is transformed into the corresponding state space expression.

Thus, the state of time k can be set as

$$x_i(k) = x_{1i}(k), (i = 1, 2, \ldots, 6).$$ (2.7)

Then the state at time $k + 1$ is

$$x_i(k + 1) = x_{1i}(k) + w_i(k), (i = 1, 2, \ldots, 6).$$ (2.8)

where $w_i(k)(i = 1, 2, \ldots, 6)$ is a Gaussian white noise sequence with zero mean value, and the state equation of the system can be obtained as

$$X(k + 1) = X(k) + W(k).$$ (2.9)

where $W(k)$ is the noise vector composed of $w_i(k)(i = 1, 2, \ldots, 6)$, and

$$E[W(k)W^{T}(j)] = Q_k\delta_{kj}.$$ (2.10)

At the same time, the observation equation of the system can be obtained based on Eq. (2.6)

$$Y(k) = H(k)X(k) + V(k),$$ (2.11)

where $V(k)$ is a Gaussian white noise sequence with zero mean and R_k covariance matrix, and $V(k)$ is independent of $W(k)$.

According to the state equation and observation equation of Buck converter system, the following recursive equation of Kalman filter can be used for parameter identification.

$$\hat{X}(k|k - 1) = \Phi(k, k - 1)\hat{X}(k - 1|k - 1)$$ (2.12)

$$P(k|k - 1) = \Phi(k, k - 1)P(k - 1|k - 1)\Phi^{T}(k, k - 1)$$
$$+ \Gamma(k, k - 1)Q_{k-1}\Gamma^{T}(k, k - 1)$$ (2.13)

$$K(k) = P(k|k-1)H^{T}(k)[H(k)P(k|k-1)H^{T}(k) + R_k]^{-1} \qquad (2.14)$$

$$\varepsilon(k) = Y(k) - H(k)\hat{X}(k|k-1) \qquad (2.15)$$

$$\hat{X}(k|k) = \hat{X}(k|k-1) + K(k)\varepsilon(k) \qquad (2.16)$$

$$P(k|k) = [I - K(k)H(k)]P(k|k-1) \qquad (2.17)$$

Then the parameter identification result is

$$\hat{X} = \begin{bmatrix} \hat{x}_{11} & \hat{x}_{12} & \hat{x}_{13} & \hat{x}_{14} & \hat{x}_{15} & \hat{x}_{16} \end{bmatrix}^{T}. \qquad (2.18)$$

The estimated values of the obtained parameter matrix are used to calculate the estimated values of the Buck converter's components, the following equations can be obtained:

$$\hat{L} = \frac{E \cdot T}{\hat{x}_{15}}, \qquad (2.19)$$

$$\hat{R} = \frac{\hat{x}_{12} \cdot E}{E - \hat{x}_{14} \cdot E - \hat{x}_{16}}, \qquad (2.20)$$

$$\hat{R}_c = \frac{\hat{x}_{16} \cdot \hat{R} \cdot \hat{L}}{\hat{R} \cdot E \cdot T - \hat{x}_{16} \cdot \hat{L}}, \qquad (2.21)$$

$$\hat{C} = \frac{\hat{L} \cdot \hat{x}_{16}}{E \cdot \hat{x}_{12} \cdot \hat{R}_c}. \qquad (2.22)$$

Thus, the parameter identification of Buck converter can be realized.

For the increase in equivalent resistor caused by the degradation of the electrolytic capacitor, it is generally regarded as a fault [3]. At the same time, it is assumed that only a single fault exists in Buck converter system at one time. The IKF algorithm is used to diagnose the Buck converter's fault in the fault state. First of all, it is necessary to establish an accurate Buck converter model and set different parameters of the models in the fault-free and faulty states for Simulink simulations. The inductor current and output voltage of Buck converter obtained by simulations are regarded as the known parameters of the system, and the parameter matrix formed by the parameter values of fault components which need to be diagnosed and other circuit components is regarded as the unknown state of the system. By constructing the IKF state equation, the parameter values of Buck converter's components at different time points are identified by using the Kalman filter recursive equations. After analyzing and comparing the parameter values of the components at each time point, the fault components and fault values of Buck converter are obtained. Thus, the fault diagnosis of Buck converter is realized.

2.3 Application Study

Based on the hybrid system model of Buck converter and the IKF based fault diagnosis method described in the chapter, the parameter identification and fault diagnosis of Buck converter are carried out. The whole experimental process is based on MATLAB2016 simulation environment, using Intel's fourth-generation core quad-core processor on Lenovo 90CYCTO1WW computer.

Based on the equivalent schematic diagram of non-ideal Buck converter, the simulation model of Buck converter built in Simulink module is shown in Fig. 2.3 and the parameters are given in Table 2.1.

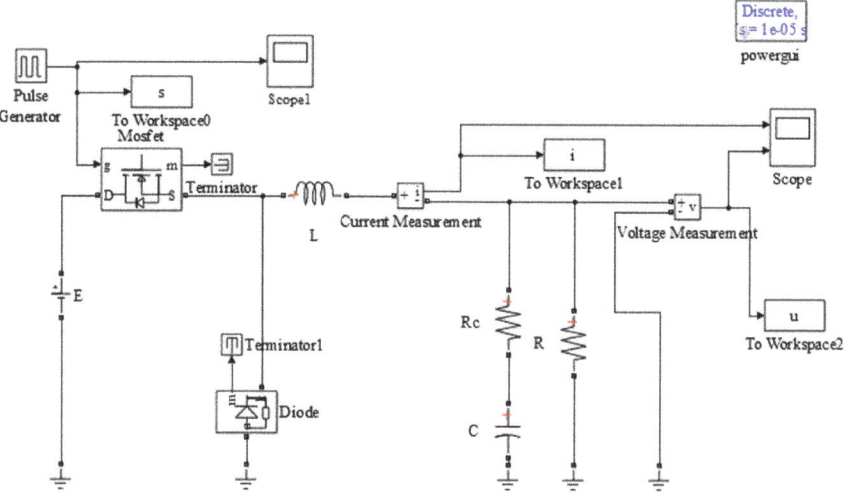

Fig. 2.3 Simulation model of Buck converter

Table 2.1 Parameter values of simulation model of Buck converter

Physical variable	True value
Input voltage E	50 V
Capacitor C	144.3 μF
Inductor L	292 μH
Working frequency f	50 kHz
Duty ratio D	0.5
Resistor R	5.76 Ω
Equivalent resistor R_c	0.46 Ω

Table 2.2 Parameter identification results of Buck converter based on IKF algorithm without fault

Number of samples	$2 \cdot L/\mu H$	$2 \cdot R/\Omega$	$2 \cdot R_c/\Omega$	$2 \cdot C/\mu F$
100	405.51	5.31	0.4507	136.45
200	349.11	5.59	0.4566	142.20
300	331.32	5.66	0.4567	143.39
400	321.97	5.64	0.4574	143.58
500	316.19	5.65	0.4579	143.85
1000	303.95	5.71	0.4596	144.87
1500	300.12	5.75	0.4593	145.36
2000	297.96	5.73	0.4600	145.31
2500	297.64	5.73	0.4581	145.30
3000	296.46	5.74	0.4590	145.45
True value	292.00	5.76	0.4600	144.30
Error (%)	1.53	0.35	0.22	0.80

Buck converter model is simulated in the Simulink interface according to the parameters. Set the simulation time as 0.03 s and the sampling time T_s as 0.00001 s, 3000 groups of inductor current and output voltage waveforms and values of Buck converter are obtained. Using the obtained values for parameter identification of IKF, the estimated results and errors of components are shown in Table 2.2.

As can be seen from Table 2.2, with the increase of the number of samples, the values of L, R, R_c and C gradually tend to the true values of components, and the errors between the final identification values of components and the true values of components are small, which indicates that the IKF algorithm is suitable for parameter identification of Buck converter in fault-free state, and the accuracy of parameter identification results is high.

It can be seen from Fig. 2.4 that identification error curves change rapidly when the number of samples changes from 0 to 1000. When the number of samples reaches 1000, the identification errors of all components has decreased to less than 5%, and then gradually tend to be stable.

Based on the same simulation conditions, the IKF algorithm and the recursive least squares (RLS for short) algorithm are compared, and the parameter identification results and errors of the Buck converter in the fault-free state are shown in Table 2.3.

From the analysis of Table 2.3, the inductor error in the identification result of the IKF algorithm is slightly larger than that of the RLS algorithm for Buck converter in the fault-free state, but the accuracy of the capacitor identification results of the IKF algorithm is significantly higher compared with the RLS algorithm, indicating that the IKF algorithm is more sensitive to changes in capacitor, and is more suitable for the fault diagnosis of the degradation of electrolytic capacitor in Buck converter studied in this chapter.

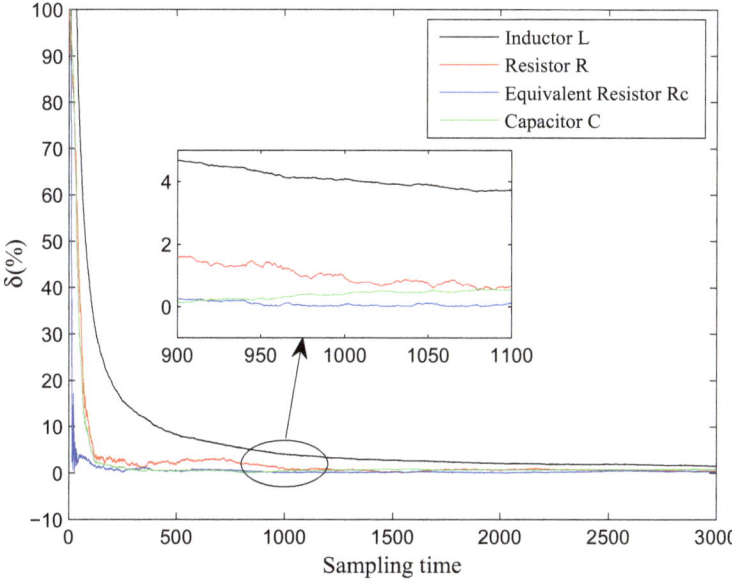

Fig. 2.4 Parameter identification errors of Buck converter elements based on IKF algorithm without fault

Table 2.3 Comparison between two algorithms in parameter identification results without fault

Algorithm	Number of samples	$L/\mu H$	R/Ω	R_c/Ω	$C/\mu F$
IKF	200	349.11	5.59	0.4566	142.20
	500	316.19	5.65	0.4579	143.85
	1000	303.95	5.71	0.4596	144.87
	2000	297.96	5.73	0.4600	145.31
	3000	296.46	5.74	0.4590	145.45
	True value	292.00	5.76	0.4600	144.30
	Error (%)	1.53	0.35	0.22	0.80
RLS	200	294.99	5.71	0.4535	145.97
	500	293.69	5.67	0.4567	145.59
	1000	292.56	5.72	0.4590	146.00
	2000	292.22	5.73	0.4597	146.14
	3000	292.60	5.74	0.4588	146.18
	True value	292.00	5.76	0.4600	144.30
	Error (%)	0.21	0.35	0.26	1.30

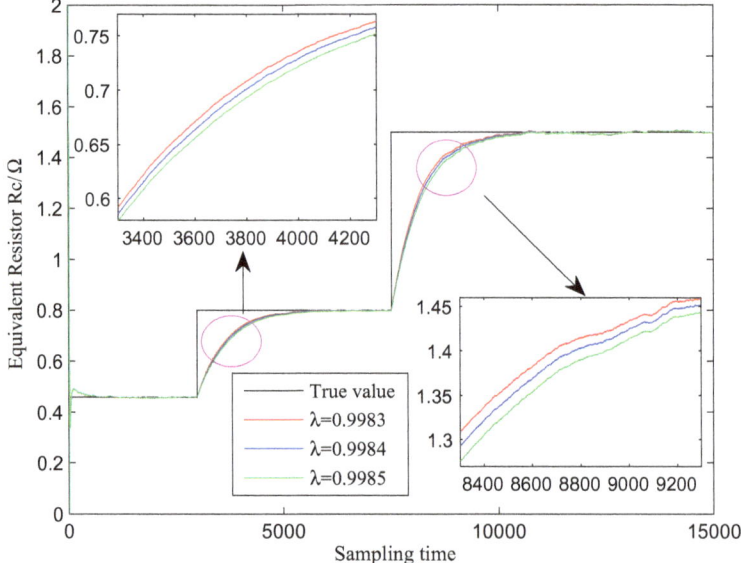

Fig. 2.5 Identification curves of equivalent resistor based on IKF algorithm in fault state

Next, the applicability of this algorithm to the electrolytic capacitor fault of Buck converters is discussed. It is assumed that the fault caused by the degradation of the electrolytic capacitor in Buck converter is shown as a sudden change of R_c. Specifically, the value of R_c is set to change from 0.46 Ω to 0.8 Ω when the number of samples is 3000, and then from 0.8 Ω to 1.5 Ω when the number of samples is 7500. The simulation time is set to 0.15 s, and other simulation experimental conditions and component parameters are the same as those in the above-mentioned fault-free state. The parameter identification curves of the equivalent resistor R_c under different forgetting factors are obtained as shown in Fig. 2.5.

Analysis of Fig. 2.5 shows that the identification curves of R_c based on the IKF algorithm under different forgetting factors can better track the true change curve of R_c. And with the decrease of forgetting factor, the degree of curve fitting is better. At the same time, considering that the degradation of the electrolytic capacitor will have a greater impact on the capacitor, the capacitor change is also analyzed below when the equivalent resistor changes. The parameter identification values of equivalent resistor and the identification errors of capacitor based on different forgetting factors are obtained as shown in Table 2.4, and the corresponding identification error curves of capacitor are shown in Fig. 2.6.

It can be seen from Table 2.4 and Fig. 2.6 that the change of equivalent resistor has a certain effect on the value of the capacitor, but the effect is small. The stable values of the identification errors of capacitor under different forgetting factors are all less than 5%, which has no effect on the fault diagnosis results of Buck converter. At the same time, the more accumulated step size, the greater the effect of

Table 2.4 Identification values of equivalent resistor and identification errors of capacitor based on IKF algorithm in fault state

Number of samples	R_c / Ω			δ_C (%)		
	$\lambda = 0.9983$	$\lambda = 0.9984$	$\lambda = 0.9985$	$\lambda = 0.9983$	$\lambda = 0.9984$	$\lambda = 0.9985$
1000	0.4639	0.4639	0.4640	4.9180	5.0082	5.1017
2000	0.4576	0.4577	0.4578	3.9665	4.0086	4.0553
3000	0.4621	0.4620	0.4619	3.4738	3.5027	3.5348
4000	0.7358	0.7292	0.7219	2.5228	2.5460	2.5752
6000	0.7981	0.7974	0.7965	3.2604	3.2367	3.2142
7500	0.7978	0.7978	0.7977	3.2700	3.2522	3.2350
10000	1.4892	1.4859	1.4817	0.6894	1.1623	1.3439
15000	1.5002	1.5005	1.5007	0.6581	1.1923	1.4184

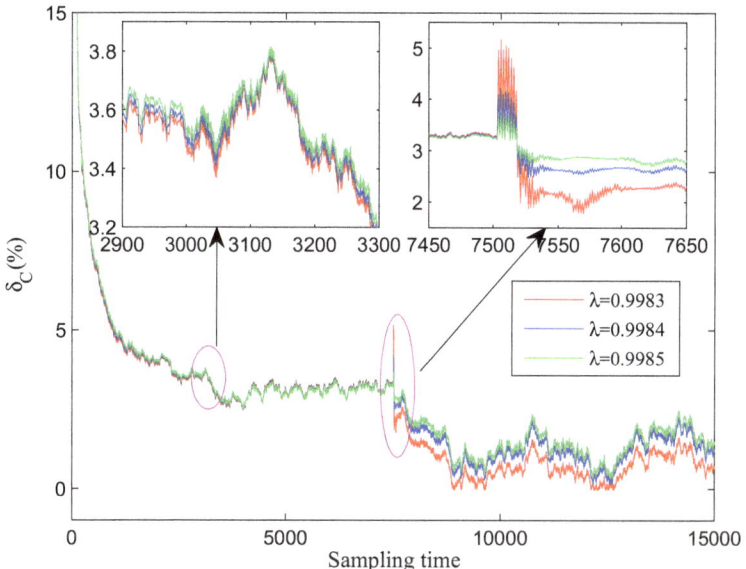

Fig. 2.6 Curves of capacitor identification errors based on IKF algorithm in fault state

historical data on the identification curve. Therefore, the curve fluctuates little when the number of samples is 3000 for the first fault point in Fig. 2.6, while the curve fluctuates violently when the number of samples is 7000 for the second fault point. In addition, according to the curves in Figs. 2.5 and 2.6, different forgetting factors will affect the followability and stability of the fault diagnosis curves. When the forgetting factor is 0.9983, the identification curve of equivalent resistor has the best tracking performance. However, the identification error curve of capacitor under this forgetting factor fluctuates more violently than the other two curves when the number

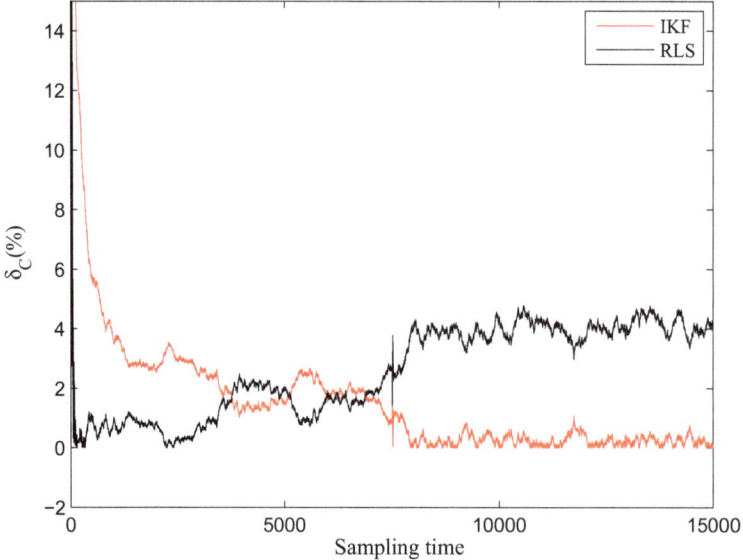

Fig. 2.7 Comparison between two algorithms in capacitor identification errors in fault state

of samples is around 7500. But meanwhile, it can be seen that the identification error curves of capacitor under different forgetting factors fluctuate basically the same after the number of samples is 7550, and when the forgetting factor is 0.9983, the identification error of capacitor is smaller than the other two cases. Therefore, when the forgetting factor is 0.9983, the fault diagnosis result is the best.

Considering that the degradation of electrolytic capacitor will have a certain impact on the size of capacitor, the IKF algorithm and RLS algorithm are used for fault diagnosis of Buck converter in fault state. The identification error curves of capacitor are shown in Fig. 2.7, and the forgetting factor in both algorithms is 0.9983.

It can be seen from Fig. 2.7 that the identification errors of capacitor of the IKF algorithm are larger than that of the RLS algorithm when the number of samples is small, but with the increase of the number of samples, the capacitor error gradually decreases and tends to zero. The capacitor identification error of the RLS algorithm is gradually increasing, so it will inevitably have an adverse effect on the system's fault diagnosis results with the increase of the number of samples. Therefore, the IKF algorithm is effective for the degradation of electrolytic capacitor.

2.4 Concluding Remarks

Aiming at the problem of parameter identification and fault diagnosis of power converter, the IKF based fault diagnosis algorithm which reverses the traditional Kalman filter recursive process is proposed in this chapter. The known matrix is constructed by using current and voltage data, and the parameters of circuit components are taken as unknowns, and the Kalman recursive algorithm is derived reversely. Simulation results and analyses show that the algorithm can accurately identify the parameters and diagnose the fault of power converter's components, and has the characteristics of strong tracking, high accuracy and good real-time performance. At the same time, compared with the RLS algorithm, the IKF algorithm has the advantages of higher accuracy and better adaptability in fault diagnosis of the electrolytic capacitor in power converter.

The IKF algorithm proposed in this chapter can not only be applied to power converter's fault diagnosis caused by the degradation of electrolytic capacitor, but also can realize real-time monitoring of the circuit in the abnormal state of the circuit topology such as inductor fault. Besides, it can also be extended to the fault diagnosis and parameter identification of the DC chopper converters and switching power supplies, such as Boost, Cuk, Sepic, Zeta, etc.

References

1. G.J. Xie, H.F. Xu, Modeling of current programmed mode non-ideal buck converter systems. Proc. CSEE **32**, 52–58 (2012)
2. Y.J. Ma, L. Ma, X.S. Zhou, Modeling and simulation of non-ideal buck converter in CCM mode based on state-space averaging method. J. Tianjin Univ. Technol. **30**(5), 13–16 (2014)
3. J. Hannonen, J. Honkanen, J.P. Strom, Capacitor aging detection in a DC-DC converter output stage. IEEE Trans. Ind. Appl. **52**(4), 3224–3233 (2016)

Chapter 3
Design of Ellipsoid Set-Membership Based Fault Detection Filter

3.1 Preliminaries and Problem Formulation

Consider a SISO dynamic system as follows:

$$y_k = \Phi_k^{\mathrm{T}} \theta + v_k; k = 1, 2, \ldots, N, \tag{3.1}$$

where $\{y_k\} \in \mathbb{R}$, $\theta \in \mathbb{R}^m$, $\Phi_k \in \mathbb{R}^m$ are the output observation sequence, unknown parameter vector and observable data vector, respectively. $\{v_k\} \in \mathbb{R}$ is a unknown but bounded measurement noise sequence that satisfies $v_k^2 \leqslant \sigma^2$. The parameter vector and the system information vector are defined by

$$\theta = \left[a_1, \ldots, a_{n_a}, b_0, \ldots, b_{n_b} \right],$$
$$\Phi_k = \left[-y_{k-1}, \ldots, -y_{k-n_a}, u_k, u_{k-1}, \ldots, u_{k-n_b} \right],$$

where $m = n_a + n_b + 1$. $\{y_k, \Phi_k\}$ is the measurement sequence pair, and v_k is bounded, i.e., $\|v_k\| \leqslant \sigma$ where $\sigma > 0$. Therefore, for the kth measurement sequence pair $\{y_k, \Phi_k\}$, the following measurement set can be obtained by

$$S(k) = \{\theta \in \mathbb{R}^m : \left| y_k - \Phi_k^{\mathrm{T}} \theta \right| \leqslant \sigma\}. \tag{3.2}$$

From Eq. (3.2), It can be get that the feasible estimation parameters lie between two hyperplanes parallel to each other. In a given time, as the pairs of measurement sequences increase, these local membership sets eventually form a convex polytope set $\Theta(k)$ that contains all feasible parameters θ, i.e.,

$$\Theta(k) = \bigcap_{i=1}^{k} S(i) = \bigcap_{i=1}^{k} \{\theta \in \mathbb{R}^m : \left| y_i - \Phi_i^{\mathrm{T}} \theta \right| \leqslant \sigma\}, \tag{3.3}$$

Z. Wang et al., *Advances in Fault Detection and Diagnosis Using Filtering Analysis*, https://doi.org/10.1007/978-981-16-5959-1_3

where $\Theta(k)$ is a convex polytope formed by the intersection of set $S(1), \ldots, S(k)$. Considering that the shape of the convex polytope is complex and irregular polyhedron or even non-convex, so it is difficult to describe the outside of the polytope. Here, the objective is to find a geometry that is easy to describe and makes it possible to include all valid arguments for a feasible set. Ellipsoid is used to approximate a feasible set of parameters because of its easily describing shape. The solution consists on recursively determining a sequence of ellipsoids $E(k)$ which enclose $\Theta(k)$. Define an ellipsoid $E(k)$ by [1]

$$E(k) = \left\{ \boldsymbol{\theta} : S_k = \sum_{k=1}^{N} \rho_k (y_k - \Phi_k^T \boldsymbol{\theta})^2 \leqslant \sum_{k=1}^{N} \rho_k \sigma^2 \right\}, \tag{3.4}$$

The tightening description is in the form of

$$E(k) = \{ \boldsymbol{\theta} : (\boldsymbol{\theta} - \boldsymbol{\theta}_c(N))^T P_N^{-1} (\boldsymbol{\theta} - \boldsymbol{\theta}_c(N)) \leqslant 1 \}, \tag{3.5}$$

where $\boldsymbol{\theta}_c(N)$ is the center of the collection for the ellipsoid, $P_N \in \mathbb{R}^{m \times m}$ is a symmetric positive definite matrix describing the shape, the size and position of an ellipsoid, $\boldsymbol{\theta}_c(N)$ is equivalent to the exception of $\boldsymbol{\theta}$, P_N is equivalent to the variance of $\boldsymbol{\theta}$. The size of P_N directly determines the size of the ellipsoid and the estimation accuracy of $\boldsymbol{\theta}$. For P_N determines the position, size and shape of the ellipsoid, in order to improve the approximation of the parameter estimates and the convergency to the true value, the P_N should be determined based on certain criterions.

3.2 Process of Ellipsoid Set-Membership Method

Accurate parameter estimation results have always been a hot spot pursued by the industry, and experts are committed to in-depth research on the minimization ellipsoid. Liang et al. [2] introduced the conditions of noise and regression vector by considering the consistency and convergence of the dynamic system.

(i) The noise sequence $\{v_k, F_k\}$ is a martingale difference sequence, where $\{F_k\}$ is a non-decreasing sub-σ-algebraic sequence, i.e., $F_k = \sigma(y_k, \Phi_k, \ldots, y_0, \Phi_0)$, F_0 contains all initial value information. There exists a constant $r > 2$ such that

$$\underset{k}{sup} \, E\left[|e_k|^r \, |F_{k-1}| \right] \leqslant \sigma^r < \infty, \tag{3.6}$$

(ii) Regression vector sequence $\{\Phi_k, F_k\}$ is any adaptive sequence, i.e., $\Phi_k \in F_k$, $\forall k \geqslant 0$.

For the regression model, if the above two conditions are satisfied and when $k \to \infty$, we can draw the following conclusions:

$$\triangle \boldsymbol{\theta}_k^T P_k^{-1} \triangle \boldsymbol{\theta}_k = O(\log r_k). \tag{3.7}$$

$$\sum_{i=0}^{k} \frac{(\Phi^{T}\triangle\theta_{k})^{2}}{1 + \Phi_{k}^{T}P_{k}\Phi_{k}} = O(\log r_{k}), \tag{3.8}$$

where $\triangle\theta_{k}$ is defined as $\triangle\theta_{k} := \theta - \hat{\theta}_{k}$ and "O" is a fixed constant

$$r_{k} = 1 + \sum_{i=1}^{k} ||\Phi_{i}||^{2}. \tag{3.9}$$

The error is only related to $O(\log r_{k})$ when the system satisfies certain conditions. The detailed description can be found in the method proposed by Liang et al. [2]. But it is different to verify whether the system meets that conditions. In addition, when the above conditions are not satisfied, the least squares algorithm must be modified. Therefore, the quadratic index weighted by sequence $\{\rho_{i}\}$ is used in this chapter:

$$J_{k}(\theta) = \sum_{i=1}^{k} \rho_{i}(y_{i+1} - \Phi_{i}^{T}\theta)^{2}. \tag{3.10}$$

In order to make the weight value ellipsoid (WVE for short) identification algorithm has a good asymptotic properties and self-convergence, define

$$\rho_{k} = \frac{1}{\log r_{k}}, \tag{3.11}$$

$$r_{k} = 1 + \sum_{i=1}^{k} ||\Phi_{i}||^{2}. \tag{3.12}$$

By calculating r_{k} in Eq. (3.12), the size and position of the ellipsoid is determined at each step and the WVE algorithm is accomplished.

3.3 Finite Data Window Algorithm

It can be seen that ρ_{k} is a non-additive sequence From the Eqs. (3.11) and (3.12), which shows the caution of the proposed WVE algorithm. However, over time and the accumulation of data, the value of the weight function ρ_{k} in Eq. (3.11) approximates zero, the tracking ability of the algorithm becomes worse and the computational complexity of the system identification algorithm increases.

Besides, the WVE algorithm uses the observation pairs from time 1 to k and the dimension of r_{k} increases with each recursion of the algorithm, as shown in Eq. (3.13),

$$r_{k+1} = 1 + \sum_{i=1}^{k+1} ||\Phi_i||^2 = r_k + ||\Phi_{k+1}||^2. \tag{3.13}$$

From Sect. 3.2, it can be get that the WVE algorithm as the disadvantage of double counting, causing low data utilization rate and poor recursive effect. Thus, for calculating the weight function ρ_k, more directly, for choosing r_k, this chapter proposes a finite data in a rolling data window with fixed data length from time t to time $t + N$, i.e., $\{\Phi_k, t \leqslant k \leqslant t + N\}$. Thus, r_k is rewritten by the following formulas,

$$r_k = 1 + \sum_{k}^{k+N} ||\Phi_i||^2, \tag{3.14}$$

$$\rho_k = \frac{1}{\log r_k}. \tag{3.15}$$

where N is defined as the data window length. The weight value ellipsoid with finite data window (WVE-FDW for short) algorithm is proposed in this chapter. The novel algorithm has the metabolic performance due to the rolling data window, that is, a new data added and the most primitive data moved in each iteration, that makes the data window length is fixed in each iteration and data utilization is also improved.

Minimization the cost function $J_k(\boldsymbol{\theta})$ with respect to $\boldsymbol{\theta}$ leads to the recursion formulas

$$\boldsymbol{\theta}_{k+1} = \boldsymbol{\theta}_k + \alpha_k P_k \Phi_k (y_{k+1} - \Phi_k^{\mathrm{T}} \boldsymbol{\theta}_k), \tag{3.16}$$

$$P_{k+1} = P_k - \alpha_k P_k \Phi_k \Phi_k^{\mathrm{T}} P_k, \tag{3.17}$$

$$\alpha_k = (\rho_k^{-1} + \Phi_k^{\mathrm{T}} P_k \Phi_k)^{-1}. \tag{3.18}$$

Generally, the running steps of WVE-FDW algorithm can be concluded as the prediction step and the correction step. By a finite number of iterations, a minimized ellipsoid is obtained and its center is considered as the parameter estimate. The process of correction as shown in Fig. 3.1 and the iteration process is listed as follows:

Iteration 1: Assume that the set of feasible states $E(k)$ has been obtained after kth calculations, i.e., $E(k) \supset \boldsymbol{\theta}(k) = \bigcap_{i=1}^{k} S(i)$. $E(k)$ begins from $E(1)$;

Iteration 2: The feasible set of feasible states $E(2)$ can be obtained from the intersection between the ellipsoid $E(1)$ and measurement set $S(2)$, i.e., $E(2) \supset E(1) \cap S(2)$; the iteration goes repeatedly with the time increases until $t = L$;

Iteration L: After Lth iterations, a minimization ellipsoid is obtained and the center of the ellipsoid $\boldsymbol{\theta}(L)$ is considered as the final parameter estimate.

Generally speaking, different data window length will lead to different identification accuracy and efficiency of the algorithm. Next, the influence of data window length on identification results is discussed.

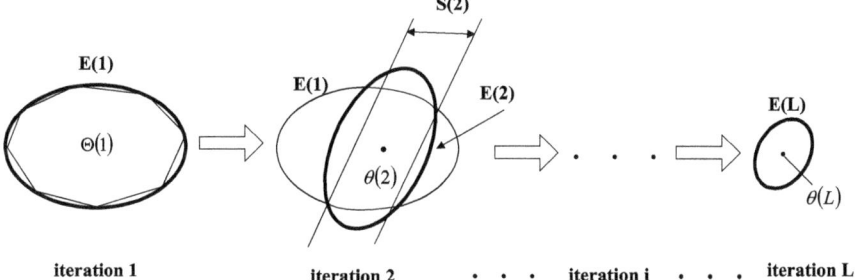

Fig. 3.1 The prediction and correction iterations of the WVE-FDW algorithm

3.4   Illustrative Simulations

Consider the following system

$$y_k = 0.8u_{k-1} + 1.35u_{k-2} + 2.25u_{k-3} + v_k. \qquad (3.19)$$

Two types of inputs are adopted to simulate separately. Here, the parameter estimation error $\delta = ||\tilde{\boldsymbol{\theta}}_k - \boldsymbol{\theta}||/||\boldsymbol{\theta}||$ ($\tilde{\boldsymbol{\theta}}_k$ is the estimate at time k) changes over time curve to illustrate the simulation results that are shown in Figs. 3.2, 3.3 and 3.4.

Case 1 :
Let $u_k \in U[-1, 1]$, $v_k \in U[-1, 1]$, where $U[a, b]$ means evenly distributed in interval $[a, b]$. The simulation results are shown in Figs. 3.2 and 3.3.

From Figs. 3.2 and 3.3, it can be seen that:

(i) The WVE and the WVE-FDW algorithm both show good convergence and high accuracy. In general, the WVE-FDW algorithm performs better.

(ii) When the measured data is large enough, the estimation error δ of both algorithms decrease steadily down to the minimum and the estimated parameters are close to the true values.

(iii) The complexity of the WVE-FDW algorithm is greatly reduced and the data utilization rate is high.

(iv) As the length of the data window increases, the convergence of parameters is better and the identification accuracy is higher.

Case 2 :
In order to demonstrate not only the proposed WVE and WVE-FDW algorithms are valid for the condition of uniformly distributed input and noise, but also are effective for the bounded trigonometric function inputs, the comprehensive analyses are implemented in this chapter. In this simulation, $u_k = \sin^2(t) + \cos(t)$ is used as the system input, the amplitude of v_k is the standard normal distribution of $\sqrt{2}$, i.e., $v_k \in \sqrt{2}N(0, 1)$, and the simulation results are shown in Fig. 3.4 and Table 3.1.

As can be seen from Fig. 3.4 and Table 3.1, when the input is a random distribution of trigonometric functions and the noise is normal, the proposed WVE-FDW

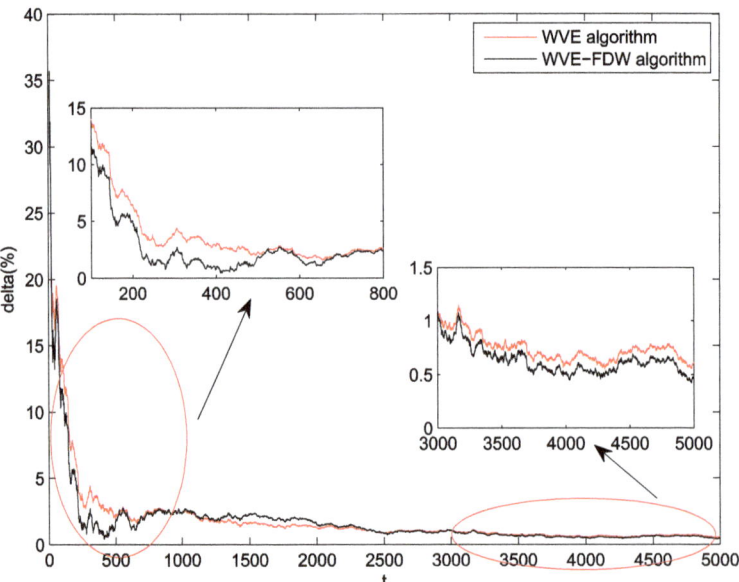

Fig. 3.2 The comparison between the WVE and WVE-FDW algorithm in estimation errors δ versus t with uniform distribution input

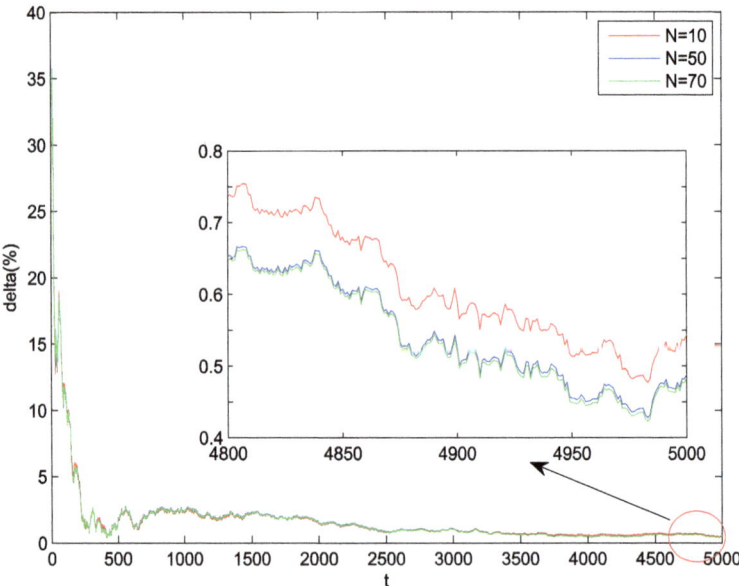

Fig. 3.3 The WVE-FDW estimation errors δ versus t on different data window lengthes with uniform distribution input

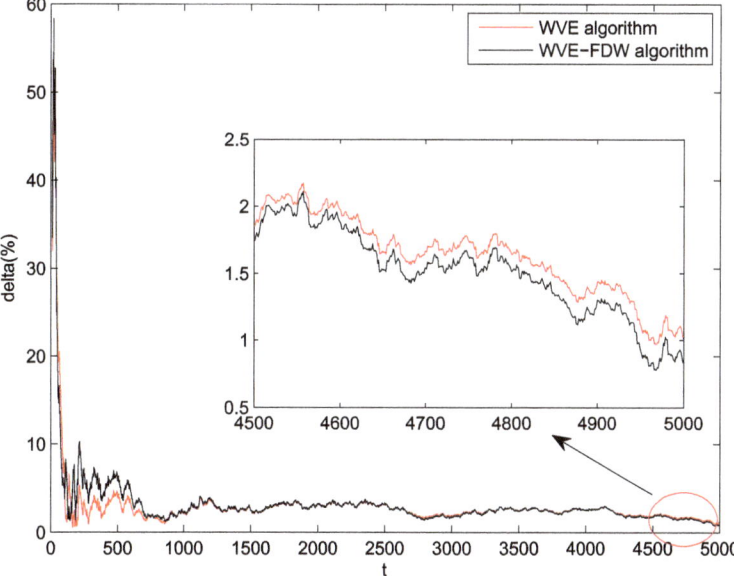

Fig. 3.4   The comparison between the WVE and WVE-FDW algorithm in estimation errors δ versus t with trigonometric function input

algorithm can still make the estimated parameters converge to the true parameters smoothly and the error of parameter estimation is small.

3.5   Concluding Remarks

In order to improve the accuracy and the dynamic adjustment of the ellipsoid estimation algorithm, a weight value ellipsoid estimation algorithm based on finite data window is proposed in this chapter. The data window fixed rolling data is used for the limited data identification under unknown noises in the proposed algorithm. After the iterations that get an ellipsoid used to approximate the feasible set of parameter, the intersection is used between observation set and the ellipsoid, and the intersection part is a valid parameter set and approximated by ellipsoid. The simulation results show that the proposed algorithm has a good accuracy and reduced computation, the identification performance becomes better as the length of data window increases.

The proposed finite data window based weight value ellipsoid estimation algorithm has some advantages in computational complexity and data utilization, so it has the characteristics of high recognition efficiency while obtaining reliable identification effect. The proposed algorithm in this chapter can also be applied to the field of state estimation [3], fault diagnosis, complex system analysis and control [4], and industrial process modeling in the future work [5].

Table 3.1 The WVE-FDW estimation errors δ versus t on different data window lengthes with trigonometric function input

N	t	a_1	a_2	a_3	$\delta(\%)$
10	200	0.72260	1.46653	2.16028	6.05831
	500	0.67993	1.40345	2.15328	5.94878
	1000	0.81069	1.29407	2.22165	2.31897
	2000	0.84605	1.27578	2.26448	3.22769
	3000	0.82894	1.31489	2.26944	1.80381
	4000	0.84514	1.31347	2.28914	2.55282
	5000	0.81906	1.34282	2.26285	0.87790
50	200	0.72320	1.46565	2.16147	6.00248
	500	0.67979	1.40410	2.15323	5.96135
	1000	0.81028	1.29374	2.22224	2.31754
	2000	0.84590	1.27590	2.26366	3.21627
	3000	0.82851	1.31531	2.26883	1.77516
	4000	0.84467	1.31357	2.28899	2.53685
	5000	0.81879	1.34314	2.26249	0.85971
70	200	0.72297	1.46592	2.16124	6.01760
	500	0.67975	1.40412	2.15323	5.96270
	1000	0.81023	1.29378	2.22221	2.31628
	2000	0.84590	1.27589	2.26366	3.21664
	3000	0.82848	1.31533	2.26882	1.77386
	4000	0.84464	1.31358	2.28898	2.53558
	5000	0.81877	1.34316	2.26248	0.85871
True value		0.80000	1.35000	2.25000	

References

1. Q. He, *The Theory and Method of Set Membership Estimation and Its Application* (Hunan University, Changsha, 2002)
2. L.M. Liang, L. Wu, Z.X. Li, An improved set-membership identification algorithm based on ellipsoid outside. Control Theory Appl. **28**(11), 11–13 (2009)
3. L. Hu, Z.D. Wang, Q.L. Han, et al., State estimation under false data injection attacks. Security analysis and system protection. Automatica **87**, 176–183 (2018)
4. H.K. Lam, A review on stability analysis of continuous-time fuzzy-model-based control systems: from membership-function-independent to membership-function-dependent analysis: Security analysis and system protection. Eng. Appl. Artif. Intell. **67**, 390–408 (2018)
5. M. Jelemensky, M. Fikar, R. Paulen, Time-optimal operation of membrane processes in the presence of fouling with set-membership parameter estimation. IFAC-PapersOnLine **50**(1), 4690–4695 (2017)

Chapter 4
Design of Polyhedron Set-Membership Based Fault Detection Filter

4.1 Preliminaries and Problem Formulation

n-dimensional (n-D) simplex: Suppose that $v_0, v_1, \ldots, v_n$ are $n + 1$ different $n - D$ vectors, if v_i-v_0 ($i = 1, 2, \ldots, n$) are linearly independent, the n-D simplex denoted by S is defined by $v_0 + \sum_{i=1}^{n} r_i(v_i - v_0)$ with $r_i \geq 0 (i = 1, 2, \ldots, n)$ and $\sum_{i=1}^{n} r_i \leq 1$. $v_0, v_1, \ldots, v_n$ are its vertices. Arbitrary n vertices form its $(n - 1)$-D facet denoted by F. In the following, S and F are described by their vertices, respectively.

n-D simplex is a special case of n-D polytope. For general n-D polytope, we can define faces of different dimension. Let us unify them by cell. A cell denoted by e_i^d represents an ith d-D cell of the polytope. We have the following cells as Table 4.1.

The definition of simplex is similar to that of cell described above. Specifically, for a simplex, every pair of vertices forms an edge and arbitrary n vertices form a $(n - 1)$-D facet.

Consider a linear discrete-time system as follow:

$$y(k) = \Psi_k^{\mathrm{T}} \theta + e(k), \quad e(k) \in [\underline{e} \ \ \bar{e}], \quad k = 1, 2, \cdots, \tag{4.1}$$

where $\theta \in \mathbb{R}^n$ is parameter vector, $y(k)$ is the system output, $\Psi_k \in \mathbb{R}^n$ is the observation vector, $e(k)$ is the noise sequence, while $\underline{e} \leqslant e(k) \leqslant \bar{e}$, $\underline{e}$ and $\bar{e}$ are known upper and lower bounds of noise $e(k)$, n is the dimension of the parameter vector. In this chapter, the feasible set of the initial parameter is characterized as a convex polyhedron. The key to the parameter identification process is to update the convex polyhedron. Equation (4.1) can be rewritten as

$$\begin{cases} y(k) - \psi_k^{\mathrm{T}} \theta \geqslant \underline{e} \\ y(k) - \psi_k^{\mathrm{T}} \theta \leqslant \bar{e}. \end{cases} \tag{4.2}$$

© The Author(s), under exclusive license to Springer Nature Singapore Pte Ltd. 2022
Z. Wang et al., *Advances in Fault Detection and Diagnosis Using Filtering Analysis*,
https://doi.org/10.1007/978-981-16-5959-1_4

Table 4.1 The simplex representation in different dimensions

e_i^0	The ith vertex of the polytope
e_i^1	The ith edge of the polytope
e_i^2	The ith 2-D polytope
$\vdots$	$\vdots$
e_i^{n-1}	The ith $(n\text{-}1)$-D facets
e_1^n	The polytope itself

Without loss of generality, in order to keep the polyhedron and the polyhedral cone are both on one side of the axis, it assumes that $\theta \geqslant 0$. If $\theta \leqslant 0$ in the actual system, the value of the parameter θ can be changed to meet the requirements by using the following two methods: One is that using variable substitution when the lower bounds of θ are known, the other one is changing the sign of the value in the observation vector. For example, consider a two-dimensional model where $b < 0$:

$$\begin{cases} y(k) = au(k) + bu(k-1) + e(k), \\ e(k) \in [\underline{e} \ \ \overline{e}], \quad k = 1, 2, \cdots, \end{cases} \tag{4.3}$$

the following transformation are made:

$$\begin{cases} y(k) = au(k) + b'(-u(k-1)) + e(k), \\ e(k) \in [\underline{e} \ \ \overline{e}], \quad k = 1, 2, \cdots, \end{cases} \tag{4.4}$$

to ensure the new parameter $b' = -b \geqslant 0$.

Considering $\theta \geqslant 0$, the solution set defined by the Eq. (4.2) can be rewritten in the matrix form

$$\begin{cases} \Phi^T \theta \geqslant b, \\ \theta \geqslant 0, \end{cases} \tag{4.5}$$

where $\Phi^T \in \mathbb{R}^{2k \times n}$ is the system information matrix, and $b \in \mathbb{R}^{2k}$ is the output vector. Because the above formula is a nonhomogeneous inequality, it can be transformed into a homogeneous inequality by adding a nonnegative parameter θ_{n+1} and shift the terms, then write it as a matrix form:

$$\begin{cases} \left[\Phi^T, -b \right] \theta^h \geqslant 0, \\ \theta^h = \begin{bmatrix} \theta_{n+1}\theta \\ \theta_{n+1} \end{bmatrix}. \end{cases} \tag{4.6}$$

Obviously, the feasible solution set of Eq. (4.5) is as same as that in Eq. (4.6). In this way, without changing the solution vector, the n-dimensional polyhedron is transformed into an $(n + 1)$-dimensional polyhedral cone.

4.2 Polyhedral Cone and the Vertices

First of all, there are several symbols that need to be introduced. A n-dimensional polyhedral cone is defined as $\{x \in \mathbb{R}^n | Ax \geqslant 0, x \geqslant 0\}$, where $A = (a_1 \quad a_2 \quad \cdots \quad a_p)^{\mathrm{T}}, a_i = [a_{i,1} \quad a_{i,2} \quad \cdots \quad a_{i,n}]^{\mathrm{T}}, i = 1, 2, \ldots, p$ are normal vectors of the polyhedral cone. It is assumed that the direction vector of each edge of the polyhedral cone is known, the polyhedral cone can also be expressed by $\{x \in \mathbb{R}^n | x = D\lambda, \lambda \geqslant 0\}$. Similarly, A n-dimensional polyhedron can be expressed by $\{x \in \mathbb{R}^n | Ax \geqslant 0, x \geqslant 0, e^{\mathrm{T}}x = 1\}$ and $\{x \in \mathbb{R}^n | x = D\lambda, \lambda \geqslant 0, e^{\mathrm{T}}\lambda = 1\}$.

It can be obtained from geometry, the basic solution set of a homogeneous inequality are the extreme rays of the polyhedral cone. A polyhedral cone is the area that satisfies the homogeneous inequalities. In order to obtain the expression for the extreme rays, we only need to solve the homogeneous equations. Assume that the length of the basic solution of the homogeneous equation corresponding to formula (4.6) is l, then the solution of (4.6) is as follows:

$$\theta^h = \theta_1^h w_1 + \theta_2^h w_2 + \cdots + \theta_l^h w_l, \quad w_i \geqslant 0, \quad i = 1, 2, \cdots, l, \tag{4.7}$$

where $\theta_j^h = \left[\theta_{1,j}^h, \theta_{2,j}^h, \cdots, \theta_{n+1,j}^h\right]^{\mathrm{T}}, j = 1, 2, \cdots, l$, is the jth solution of the homogeneous inequalities and it is also the jth extreme ray of the polyhedral cone.

Because a variable is added in the process of polyhedron becoming a polyhedral cone, that is, the $(n + 1)$th solution, from Eqs. (4.6) and (4.7), the parametric description of the feasible set of inequalities (4.5) can be written as

$$\theta_i = \frac{\theta_{i,1}^h w_1 + \theta_{i,2}^h w_2 + \cdots + \theta_{i,l}^h w_l}{\theta_{n+1,1}^h w_1 + \theta_{n+1,2}^h w_2 + \cdots + \theta_{n+1,l}^h w_l}, \quad i = 1, 2, \cdots, n. \tag{4.8}$$

Lemma 1 *if $\theta_{n+1,j}^h \neq 0$ for all $j = 1, 2, \cdots, l$, the polyhedron described by inequalities (4.5) is bounded and it can be characterized by its vertices. Especially, if we let $v^j = \left[v_1^j, v_2^j, \cdots, v_n^j\right]^{\mathrm{T}}$ be the jth vertex of the polyhedron, then the ith component of jth vertex is given by [1]:*

$$v_i^j = \frac{\theta_{i,j}^h}{\theta_{n+1,j}^h}, i = 1, 2, \cdots, n. \tag{4.9}$$

The polyhedron represented by the set of inequalities (4.5) can also be represented by vertex coordinates, specifically:

$$\theta = w_1 v^1 + w_2 v^2 + \cdots + w_l v^l, \tag{4.10}$$

where $w_i \geqslant 0$ for $i = 1, 2, \cdots, l$ and $\sum_{i=1}^{l} w_i = 1$.

The n-dimensional polyhedral cone can be expressed in the following form based on the theory of convex optimization:

$$I = \{x \in \mathbb{R}^n | Ax \geqslant 0, x \geqslant 0\}, \tag{4.11}$$

where A is the support vector matrix of each side of the polyhedral cone.

According to the Lemma 1, a polyhedral cone can also be represented by its edge vectors. Assume that I has l edges, the vector of these edges constitutes a matrix $D = (d_1, d_2, \cdots, d_l)$, and the kth column of D is the direction vector of the kth edge of I. Therefore, the representation of the polyhedral cone can also be expressed by

$$I' = \{x | x = D\lambda, \lambda \geqslant 0\}. \tag{4.12}$$

Similarly, the convex optimization of n-dimensional polyhedron is:

$$S = \{x | Ax \geqslant 0, x \geqslant 0, e^T x = 1\}. \tag{4.13}$$

Because D is the solution set of the polyhedral cone and it satisfies $e^T d_i = 1, i = 1, 2, \cdots, l$, another expression for the polyhedron is

$$S' = \{x | x = D\lambda, \lambda \geqslant 0, e^T \lambda = 1\}. \tag{4.14}$$

e is an n-dimensional unit vector in all the above formulas. The dimensions of e and λ are consistent with each expression. As can be seen from the above method, when the edge vector of the polyhedral cone is found, the vertices of the corresponding polyhedron can be known.

4.3   Multi-objective Linear Programming

Add a positive parameter θ_{n+1} to (2.1) and rewrite it as

$$f_k(\theta) = [\Phi^T, -b] \theta^h, \tag{4.15}$$

where $\theta^h = \begin{bmatrix} \theta_{n+1}\theta \\ \theta_{n+1} \end{bmatrix}$. Use it as the objective function of step k. Use Eq. (4.6) as a constraint for linear programming to form a multi-objective linear program.

In the process of solving the multi-objective problem, we first need to transform the multi-objective planning problem to obtain a single-objective optimization problem. After that, the optimal solution of the single-objective problem using nonlinear optimization algorithm is regarded as the optimal solution of the problem. By considering the importance degree of the k objective functions $f_k(\theta)$, a certain weight coefficient w is, respectively, assigned, and then all the objective functions are weighted and summed as a new objective function, and the optimal value of the new objective function is obtained in the feasible domain S of the multi-objective programming problem. In this chapter, we consider the following single-objective optimization problem:

$$\begin{cases} min \quad Z(\theta) = \sum_{i=1}^{k} w_i f_i(\theta) \\ s.t. \quad \theta \in S = \left\{ \theta \in \mathbb{R}^n | g(\theta) \geqslant 0, \theta \geqslant 0, e^{\mathrm{T}}\theta = 1 \right\}, \end{cases} \tag{4.16}$$

where $g(\theta) = \left[\Phi^{\mathrm{T}}, -b \right] \theta^h$. We take the optimal solution of the single-objective optimization problem as the optimal solution of multi-objective programming problem under a linear weighted sum. The weight vector w is taken from the following set:

$$W = [w_1 \quad w_2 \quad \cdots \quad w_k] \quad W = \{w | w_i > 0, \sum_{i=1}^{k} w_i = 1\}. \tag{4.17}$$

Remark 4.1 For any $w \in W$, the optimal solution of the single-objective optimization problem must be an effective solution of multi-objective programming problem. $\Box$

Remark 4.2 If the multi-objective programming problem is a multi-objective convex programming problem, then, for any valid solution θ^* of the multi-objective programming problem, there must be an $w \in W$, so that θ^* is the optimal solution of the single-objective optimization problem. $\Box$

We can get from the above that D are direction vectors of the polyhedral cone, and, if $e^{\mathrm{T}} D_i = e^{\mathrm{T}}$, D are vertices of polyhedron. $A = (a_1, a_2, \cdots, a_n)$ are normal vectors of the polyhedral cone. P are the direction vectors of edges of the polyhedral cone. $S_{i-1} = \{x | a_1^{\mathrm{T}} x \geqslant 0, a_2^{\mathrm{T}} x \geqslant 0, \cdots, a_{i-1}^{\mathrm{T}} x \geqslant 0, x \geqslant 0, e^{\mathrm{T}} x = 1\}$. $S'_{i-1} = \{x | x = D_{i-1}\lambda_{i-1}, \lambda_{i-1} \geqslant 0, e^{\mathrm{T}}\lambda_{i-1} = 1\}$. $S_{i-1} = S'_{i-1}$. The optimal solution is obtained by the objective function generated by each step and the nonredundant constraint condition, and the optimal solution is expanded into the optimal surface, and the intersection region with the feasible parameter domain of the FJ method is used as the final parameter feasible domain.

1. Initialization. Let $D_0 = E$ and $A_0 = E$, where E is an n-dimensional identity matrix. $S_0 = \{x | x \geqslant 0, e^{\mathrm{T}} x = 1\}$.

2. The support hyperplane normal vector a_i and the objective function f_i are calculated for each step according to the input and output sequence.
3. Determine the relationship between elements in a_i and 0, and there are three possible patterns related to a_i.

   (a) If $a_i < 0$, none of the columns of D_{i-1} satisfy the new inequality, and S_i is empty. Algorithm ends or restarts.
   (b) If $a_i > 0$, all of the columns of D_{i-1} satisfy the new inequality, and the new inequality is redundant. $D_i = D_{i-1}$, $A_i = A_{i-1}$.
   (c) Otherwise, the new inequality is not redundant. $\alpha = a_i D_{i-1}$, that is, it supports the hyperplane normal vector multiplied by the polyhedron vertices, and α and D_{i-1} have the same number of columns.

4. Determine the relationship between elements in α and 0, and there are three possible patterns related α.

   (a) If $\alpha < 0$, none of the vertices of the polyhedron satisfy the new inequality $a_i x \geqslant 0$, and S_i is empty. The algorithm ends or restarts.
   (b) If $\alpha > 0$, all the vertices of the polyhedron satisfy the new inequality $a_i x \geqslant 0$, and the supporting hyperplane normal vector formed by this step is redundant. $D_i = D_{i-1}$, $A_i = A_{i-1}$.
   (c) Otherwise, calculate the index of α whose column vectors are less than 0, and discard the column vector of the same index in D_{i-1}. $A_i = [A_{i-1}; a_i]$. Calculate all edge direction vectors P_i of the polyhedral cone by using A_i.

5. Determine whether or not P_i are all supporting hyperplanes. For example, judging the jth edge of the polyhedron gives $l = A_i P_i(:, j)$.

   (a) If $l > 0$, P_i are in all supporting hyperplanes, $P'_i = P_i$.
   (b) Otherwise, P_i are not in all supporting hyperplanes, so discard $P_i(:, j)$. Use the direction vector P'_i and $e^T x = 1$ to solve the polyhedron vertices d, $D_i = [D_{i-1}, d]$.

6. Update the supporting hyperplane normal vectors by using polyhedron vertices. For example, judging the jth normal vector of A_i gives $r = A_i(j, :)D_i$.

   (a) If $r > 0$, all vertices are within the supporting hyperplane, indicating that the hyperplane is redundant, so discard $A_i(j, :)$.
   (b) If $r < 0$, the supporting hyperplane does not contain any vertices, so discard $A_i(j, :)$.
   (c) Otherwise, the supporting hyperplane is a boundary hyperplane, so retain $A_i(j, :)$.

7. Weight multi-objectives into a single objective using Eq. (2.16). Each of these weights is a random value. The constraint is $\{x | Ax \geqslant 0, e^T x = 1\}$. Then find the optimal solution
8. Expand the optimal points into faces, where DIS is several times of the farthest distance from the center estimate point to the parameter feasible set D.

9. Transform the parameter feasible set obtained in the FJ algorithm into a parameter feasible domain (with all feasible points being wrapped with a convex hull).
10. Find the intersection of the feasible fields of the last two steps. The vertices of the intersecting feasible domain or the polyhedron vertices within the feasible domain are taken as the feasible parameter set.
11. Find the center value of the feasible parameter set as the final identification result.

4.4   Illustrative Simulations

A three-dimensional model is considered as follow:

$$y(k) = au(k) + bu(k-1) + cu(k-2) + e(k), k = 1, 2, \cdots, 65,$$

where $a = 0.50$, $b = 2.35$, $c = 1.25$. The input signal $u(k) = -1.5$ if $k = 1, 2, 3$ mod 4 while $u(k) = 1.5$ if $k = 0$ mod 4. The noise term $e(k)$ is optionally k points in a half cycle of the sinusoidal function symmetric about the origin, i.e., $\|e(k)\| \leqslant \sin(0.49\pi)$.

The projection of parameter feasible set using the FJ algorithm and the MOLPFJ algorithm are shown in the Figs. 4.1, 4.2, 4.3, 4.4, 4.5, 4.6 and 4.7. Moreover, compare the two algorithms with the minimizing the segments of the zonotope method

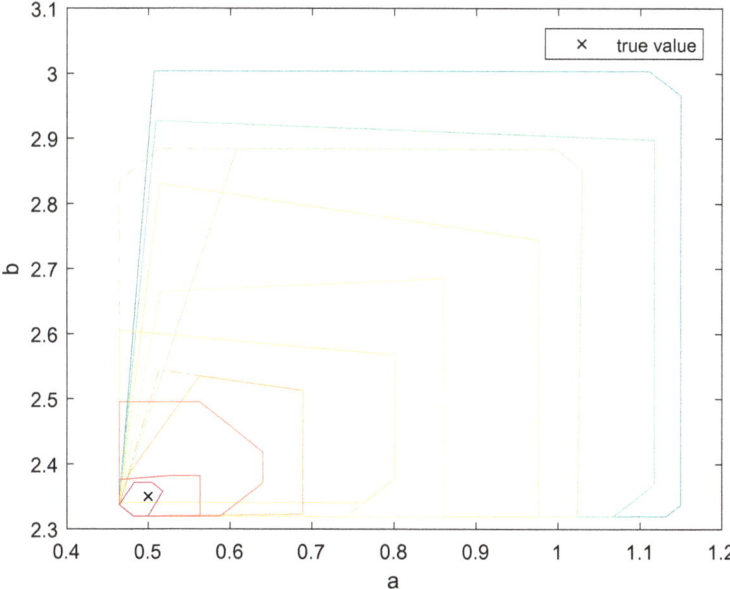

Fig. 4.1 Projection on the X-Y axis

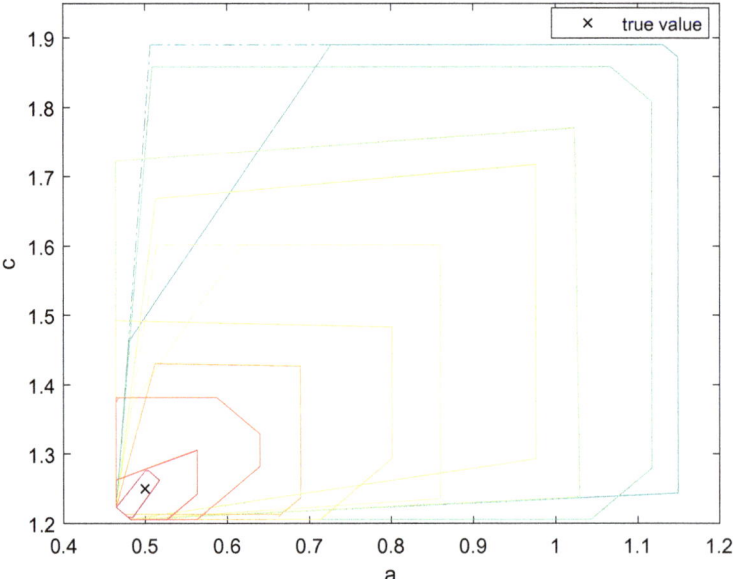

Fig. 4.2 Projection on the *X-Z* axis

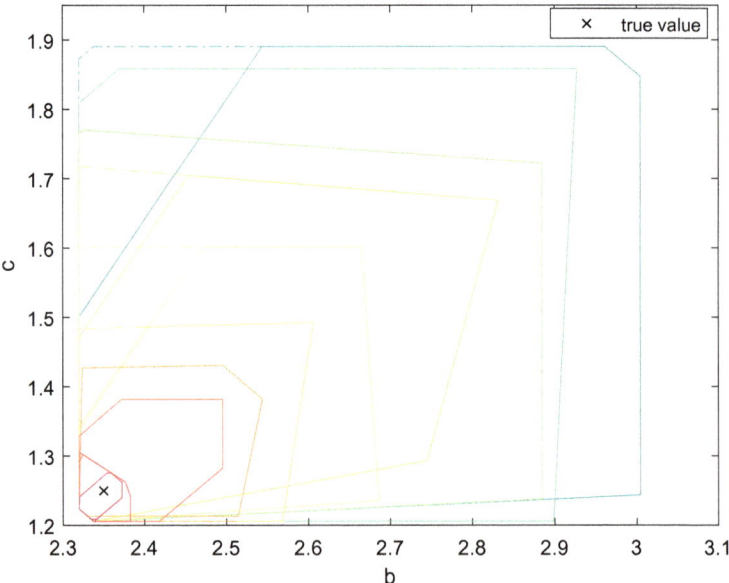

Fig. 4.3 Projection on the *Y-Z* axis

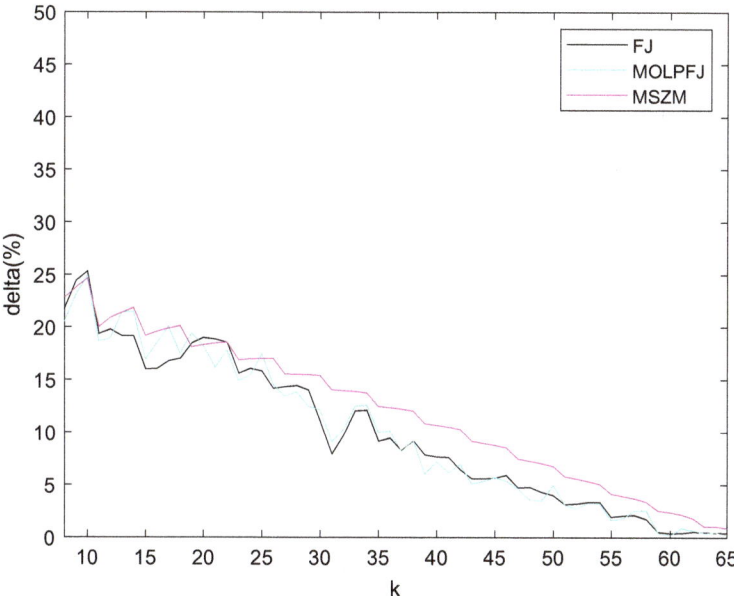

Fig. 4.4 Comparison of errors between the FJ and MOLPFJ algorithms

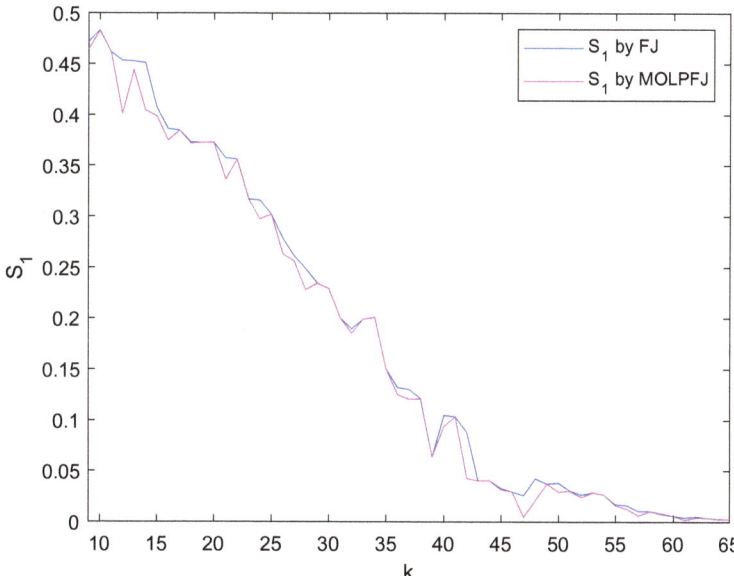

Fig. 4.5 The area of the feasible solution set on the X-Y axis

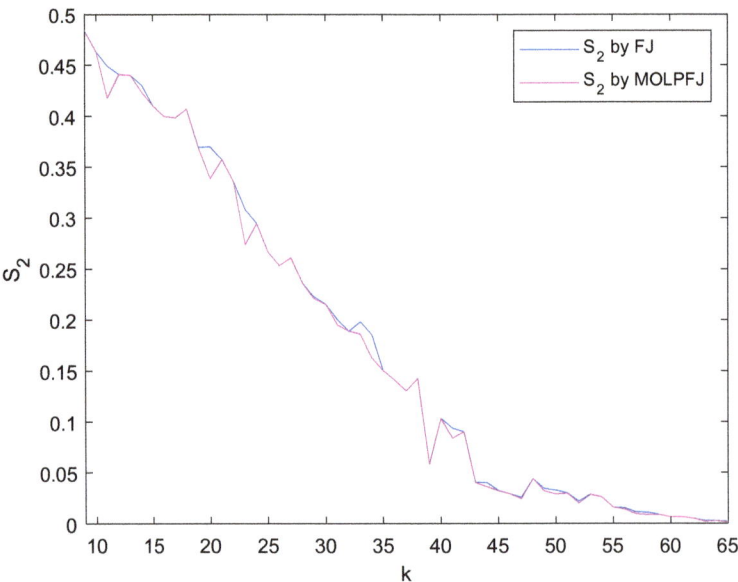

Fig. 4.6 The area of the feasible solution set on the *X-Z* axis

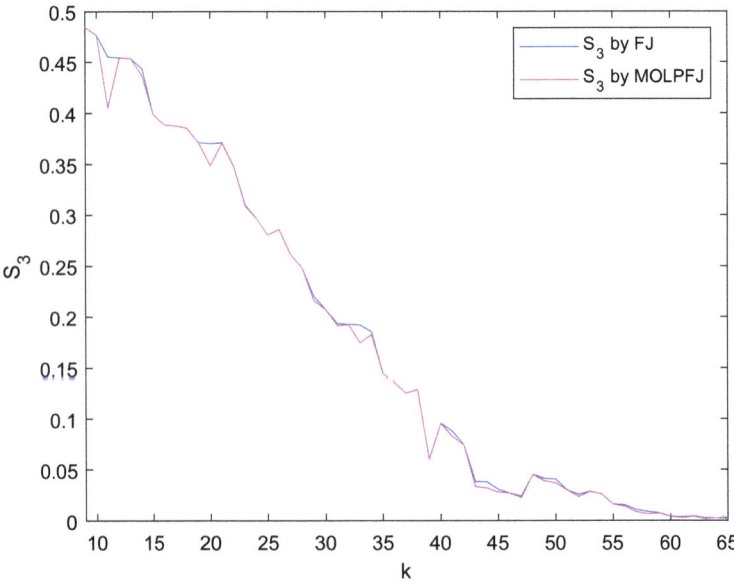

Fig. 4.7 The area of the feasible solution set on the *Y-Z* axis

(MSZM) described in [2] and the initial zonotope is $[2, 2, 2]^T \oplus 10 diag([1, 1, 1])B^3$. δ stands for absolute error, where $\delta = \sqrt{\frac{(\bar{\theta}-\theta)^2}{m\theta^2}} \times 100\%$, where $\bar{\theta}$ is the estimated value, θ is the true value and m represent number of groups identified by each parameter. S_1, S_2 and S_3 represent the area of axis X-Y, X-Z and Y-Z, respectively.

1. If $k = 0$ mod 6, the projection of three coordinate planes of MOLPFJ and FJ are shown in Figs. 4.1, 4.2 and 4.3. The areas enclosed by solid lines and dot-dash lines represent the parameter feasible regions of MOLPFJ and FJ, respectively. We can see that the areas of MOLPFJ are always smaller than that of FJ and the set of feasible parameters is decreasing and approaching the true value. The identification accuracy is continuously improved and finally reache.
2. In Fig. 4.4, MOLPFJ is the result of expanding the optimal solution obtained by multi-objective linear programming into the optimal surface and then modifying FJ. In FJ, $m = 1$. In MOLPFJ, each parameter will produce two groups of data and $m = 2$. From the numerical example, when the correction step works, the parameter feasible set area of MOLPFJ will be smaller than FJ, where can be seen in Figs. 4.1, 4.2 and 4.3, but more points far from the true value will be generated, which will increase the error. When the two methods are compared with MSZM, respectively.
3. It is shown in Figs. 4.5, 4.6 and 4.7 that the area of the projection of three coordinate planes. The black line representing the area of MOLPFJ always coincides with or is below the red line representing FJ. From the perspective of the parameter feasible set area, it can be shown that MOLPFJ is superior to FJ.

4.5 Application Study

A variable-pitch subsystem of a wind turbine are also considered in the simulation of the algorithm. The variable-pitch subsystem is an important part of the wind turbine controlling blade and pitch angle. Its mathematical model can be expressed by [3].

$$\begin{bmatrix} \dot{\beta} \\ \dot{\beta}_a \end{bmatrix} = \begin{bmatrix} 0 & 1 \\ -\omega_n^2 & -2\zeta\omega_n \end{bmatrix} \begin{bmatrix} \beta \\ \beta_a \end{bmatrix} + \begin{bmatrix} 0 \\ \omega_n^2 \end{bmatrix} \beta_r, \tag{4.18}$$

where β and β_a are the pitch angle and angular velocity, respectively, β_r is the pitch reference value, ω_n and ζ are system parameters. $\omega_n = 11.11$ rad/s, $\zeta = 0.6$ are the system natural frequency and damping coefficient, respectively. The closed-loop dynamics of the system can be approximated as the following second-order model [4]:

$$\frac{y}{u} = \frac{\omega_n^2}{s^2 + 2\zeta\omega_n s + \omega_n^2}, \tag{4.19}$$

where $y = \beta$, $u = \beta_r$, according to the literature, the sampling time is $T_s = 0.01s$. Discretizing the closed-loop system as $A(z)y(t) = B(z)u(t) + e(t)$, where $A(z) =$

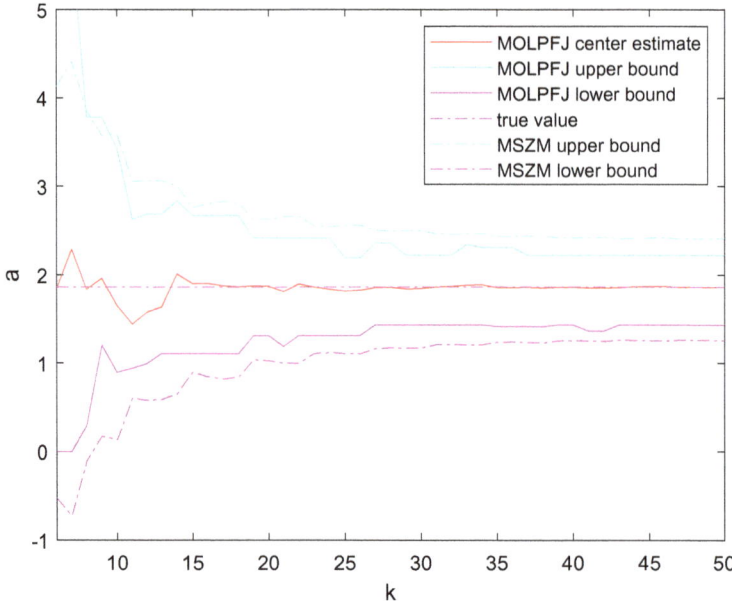

Fig. 4.8 Maximum estimated value, minimum estimated value and central estimated value of parameter a

$1 - 1.864z^{-1} + 0.8752z^{-2}$, $B(z) = 0.0059z^{-1} + 0.0056z^{-2}$, the system noise $\|e(t)\| \leqslant \gamma$ is unknown but bounded, the true value of the parameter vector is $\theta = [-1.864 \ 0.8752 \ 0.0059 \ 0.0056]^T$, and $n = 4$. The input $u(k) = 30$ when $k=1, 2, 3 \bmod 4$ while $u(k) = -30$ when $k = 0 \bmod 4$. The noise $e(k)$ is optionally k points in a half cycle of the sinusoidal function symmetric about the origin, i.e., $\|e(k)\| \leqslant \sin(0.01\pi)$. The parameter vector $[a \ b] = [-1.864 \ 0.8752]$. The error δ is defined by $\delta = \sqrt{\frac{(\bar{\theta}-\theta)^2}{\theta^2}} \times 100\%$, where $\bar{\theta}$ is the estimated value, θ is the true value. The initial zonotope of MSZM is $[2, 1, 0.01, 0.01]^T \oplus 2diag([1, 1, 1, 1])B^4$. By using the MOLPFJ algorithm and comparing with MSZM, the parameter changes as shown in Figs. 4.8, 4.9 and 4.10.

The parameter changes by using MOLPFJ and a comparison of the two methods are shown in Figs. 4.11, 4.12 and 4.13.

1. All parameters must be positive during the identification process and $\theta = [-\theta_1 \ \theta_2]$. In Figs. 4.8 and 4.9, the black solid line and black dotted line indicate the upper bound of the feasible set of parameters of MOLPFJ and MSZM, the blue solid line and blue dotted line indicate the lower bounds. We take the central value as the point estimate and we can find that the central point estimates approximate the true value with high precision. Moreover, the upper and lower bounds of MOLPFJ are tighter than MSZM, that is, it has higher accuracy.

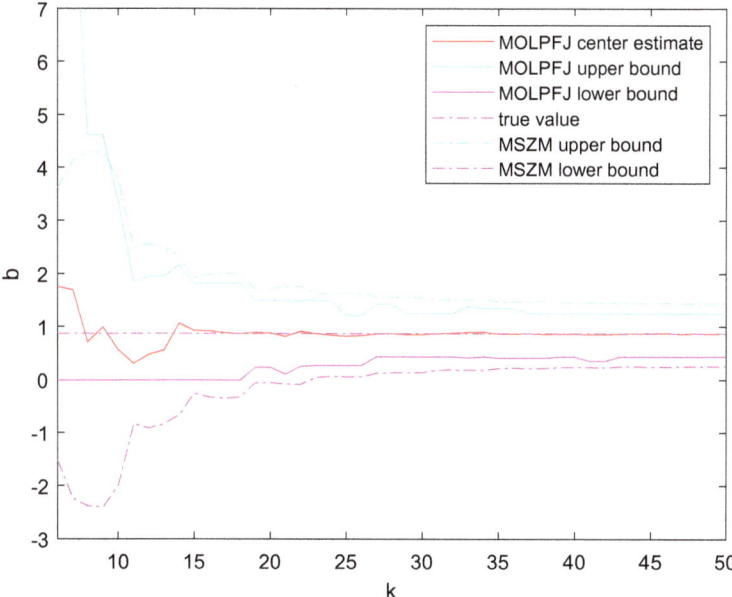

Fig. 4.9 Maximum estimated value, minimum estimated value and central estimated value of parameter a

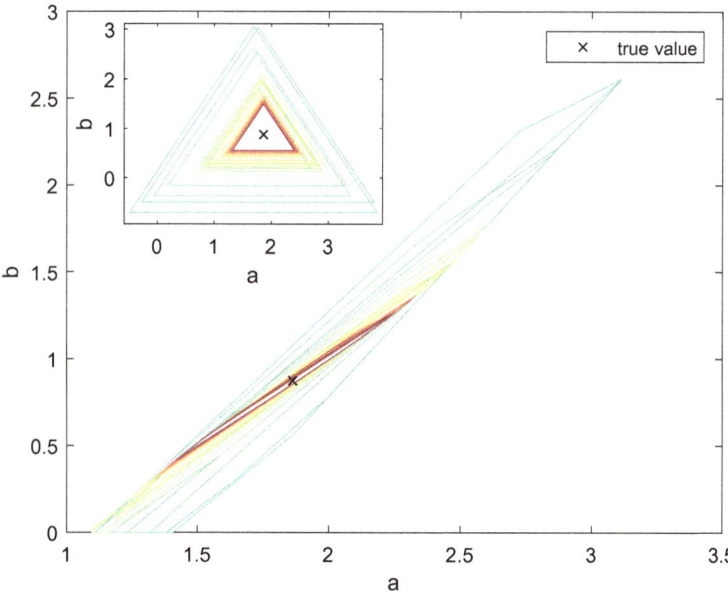

Fig. 4.10 Variation of the feasible set of parameters under two-dimensional projection

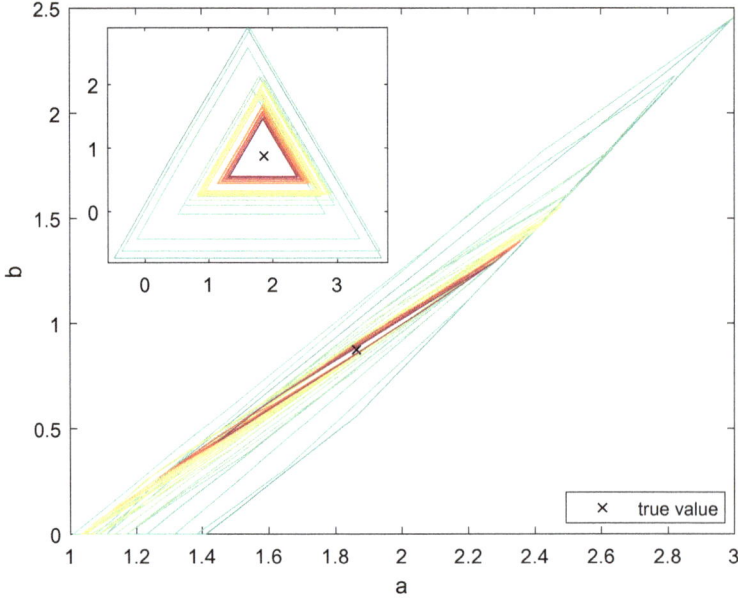

Fig. 4.11 Variation of the feasible set of parameters under two-dimensional projection

2. In Fig. 4.10, we can see the true projection of the parameter feasible set on the two-dimensional plane. Because of the flatness of the graph, it is not easy to observe the parameter set change process. After processing, we select the farthest distance from the center estimate point to the feasible set of parameters at this moment as DIS, and we convert the graph into an equilateral triangle, as shown at the bottom right in Fig. 4.10. The feasible set of parameters is decreasing but not decreasing at the end.

1. In Figs. 4.11 and 4.12, the projection of the feasible parameter set is similar to that of FJ algorithm. From the normalized graph, it can be seen that the feasible parameter set is continuously reduced and finally does not change.
2. The area of the parameter feasible set is the area of the convex hull that wraps all the parameter estimates. From Fig. 4.13, it can be seen that the areas of the two method are both decreasing, with the area of MOLPFJ always being less than or equal to the area of FJ.

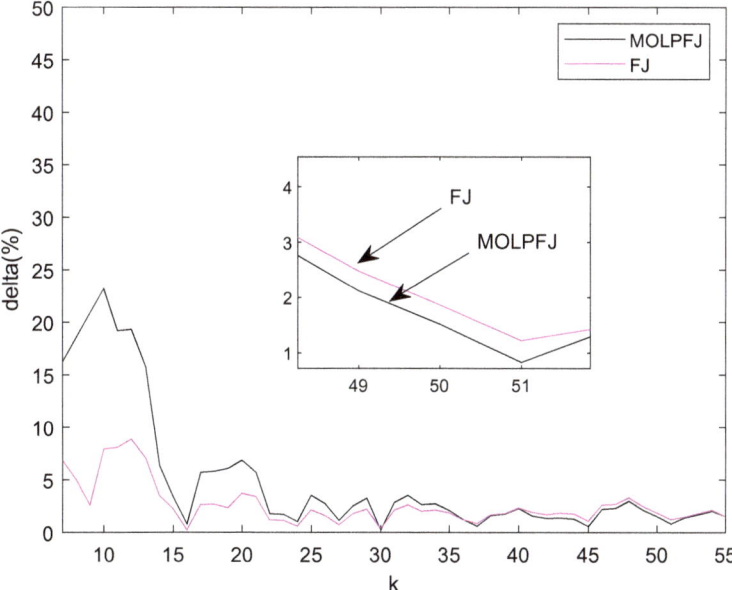

Fig. 4.12 Comparison on errors between the FJ and the MOLPFJ algorithms

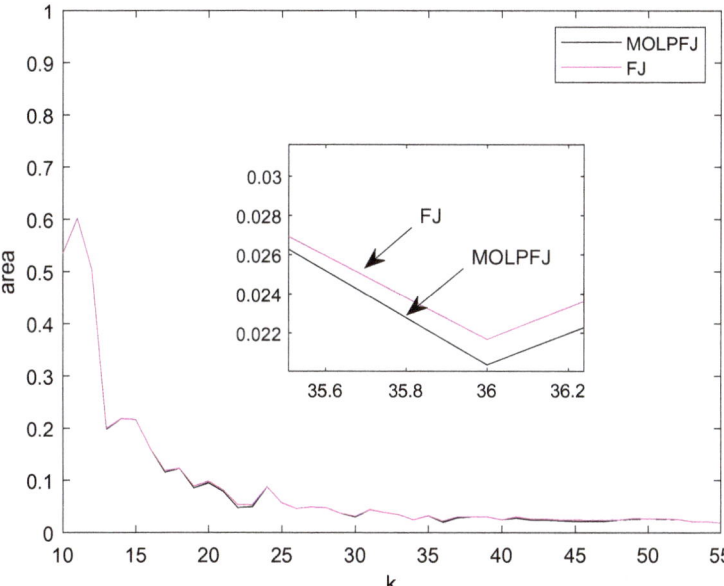

Fig. 4.13 Comparison of two-dimensional projected area of the FJ and the MOLPFJ algorithms

4.6   Concluding Remarks

Based on the increase in the dimension of one parameter, the constraints in the identification process can be transformed from inhomogeneous inequalities into homogeneous inequalities. Once the edge vectors of the convex polyhedral cone are found,
the vertices of the convex polyhedron are found. We give two methods: the FJ method
and a method combining four judgments and linear programming. Both methods are
recursive algorithms, and redundant data are eliminated during the judgment process.
We visualize the changes in the feasible set of parameters and visually observe that the
feasible set of parameters is continuously decreasing and constantly approaching the
true value. Compared to other exact algorithms, these methods have advantages that
save computational and memory cost, while also attaining higher precision. From a
comparison, it is found that MOLPFJ has higher recognition accuracy. The proposed
algorithms can also be applied to combine with the robust control algorithms [5, 6]
and to solve the estimation of switched systems [7–10] and other nonlinear models
[11–13].

References

1. Y. Wang, V. Puig, G. Cembrano, Set-membership approach and Kalman observer based on
   zonotopes for discrete-time descriptor systems. Automatica **93**, 35–43 (2018)
2. T. Alamo, J.M. Bravo, E.F. Camacho, Guaranteed state estimation by zonotope. Automatic **41**,
   1035–1043 (2005)
3. P. Casau, P. Rosa, A set-valued approach to FDI and FTC of wind turbines. IEEE Trans. Control
   Syst. Technol. **23**(1), 245–263 (2014)
4. S.M. Tabatabaeipour, Active fault detection and isolation of discrete-time linear time-varying
   system: a set-membership approach. Int. J. Syst. Sci. **46**(11), 1917–1933 (2015)
5. X.H. Chang, R. Huang, H.Q. Wang, L. Liu, *Robust Design Strategy of Quantized Feedback
   Control*. Express Briefs, IEEE Transactions on Circuits and Systems II (2019)
6. X.H. Chang, R.R. Liu, J.H. Park, *A Further Study on Output Feedback H_∞ Control for Discrete-
   Time Systems* Express Briefs, IEEE Transactions on Circuits and Systems II (2019)
7. J. Cheng, J.H. Park, J.D. Cao, and W.H. Qi. Hidden Markov Model-based nonfragile state estimation of switched neural network with probabilistic quantized outputs. IEEE Trans. Cybern.
   (2019)
8. J. Cheng, D. Zhang, W.H. Qi, J.D. Cao, and K.B. Shi. Finite-time stabilization of T–S fuzzy
   semi-Markov switching systems: a coupling memory sampled-data control approach. J. Frankl.
   Inst. (2019)
9. J. Cheng, J.H. Park, X.D. Zhao, J.D. Cao, W.H. Qi, Static output feedback control of switched
   systems with quantization: a nonhomogeneous sojourn probability approach. Int. J. Robust
   Nonlinear Control **29**(17), 5992–6005 (2019)
10. J. Cheng, Y. Zhan, Nonstationary I_2-I_∞ filtering for Markov switching repeated scalar nonlinear
    systems with randomly occurring nonlinearities. Appl. Math. Comput. **365**, 124714 (2020)
11. F. Ding, X. Zhang, L. Xu, The innovation algorithms for multivariable state-space models. Int.
    J. Adapt. Control Signal Process. **33**(11), 1601–1618 (2019)
12. Y.J. Wang, F. Ding, M.H. Wu, Recursive parameter estimation algorithm for multivariate output-
    error systems. J. Frankl. Inst. **355**(12), 5163–5181 (2018)
13. Y.J. Wang, F. Ding, Recursive least squares algorithm and gradient algorithm for Hammerstein-
    Wiener systems using the data filtering. Nonlinear Dyn. **84**(2), 1045–1053 (2016)

Chapter 5
Design of Interval Set-Membership Based Fault Detection Filter

5.1 Preliminaries and Problem Formulation

Consider the following linear time-invariant system:

$$\begin{cases} x_{k+1} = Ax_k + Bu_k + Ew_k + Ff_k, \\ y_k = Cx_k + Dv_k. \end{cases} \tag{5.1}$$

where, $x_k \in \mathbb{R}^n$ is the state vector of the system, $u_k \in \mathbb{R}^p$ and $y_k \in \mathbb{R}^q$ are the input and output vector of the system, respectively, $w_k \in \mathbb{R}^s$ and $v_k \in \mathbb{R}^l$ are unknown but bounded disturbance and noise, respectively, $f_k \in \mathbb{R}^m$ is actuator fault of any form. A, B, C, D, E and F are constant matrices with appropriate dimensions.

Regarding the actuator fault in the system (5.1) as part of the state vector, the augmented state vector is obtained:

$$\overline{x}_k = \begin{bmatrix} x_k \\ f_k \end{bmatrix}, \tag{5.2}$$

Then, construct the following augmentation system [1]:

$$\begin{cases} \overline{x}_{k+1} = \overline{A}\overline{x}_k + \overline{B}u_k + \overline{E}w_k + \overline{G}\Delta f_k, \\ y_k = \overline{C}\overline{x}_k + \overline{D}v_k. \end{cases} \tag{5.3}$$

where, $\overline{A} = \begin{bmatrix} A & F \\ 0 & I_m \end{bmatrix}, \overline{B} = \begin{bmatrix} B \\ 0 \end{bmatrix}, \overline{E} = \begin{bmatrix} E \\ 0 \end{bmatrix}, \overline{G} = \begin{bmatrix} 0 \\ I_m \end{bmatrix}, \overline{C} = \begin{bmatrix} C & 0 \end{bmatrix}, \overline{D} = D, \Delta f_k = f_{k+1} - f_k$.

From the augmented state vector (5.2), It can be seen that it is possible to observe the actuator fault f_k while estimating the state vector x_k in the augmented system (5.3). The purpose of this chapter is to design an interval observer to estimate the interval of the augmented state given the unknown but bounded disturbance and noise.

© The Author(s), under exclusive license to Springer Nature Singapore Pte Ltd. 2022   57
Z. Wang et al., *Advances in Fault Detection and Diagnosis Using Filtering Analysis*,
https://doi.org/10.1007/978-981-16-5959-1_5

The feasible set is contracted by the vector set inversion interval filtering algorithm. the compact interval estimate $[f_k^-, f_k^+]$ of the actuator failure f_k is obtained so as to realize the actuator failure observation of the system (5.1).

5.2 SIVIA Approach

SIVIA (Set Inversion Via Interval Analysis) approach [2] is to solve the nonlinear bounded errors estimation problem of feasible set based on interval analysis and be described now.

5.2.1 Interval Analysis

The idea of using intervals to represent bounded errors was first proposed in the 1950s. Moore first specifically proposed interval analysis [3] in 1966, and then many researches in this field were published [4, 5]. Some notion definitions of interval analysis are introduced.

Definition 1 An interval $[x]$ of $\mathbb{R}$ is defined as a closed, bounded and connected set $[x] = [x^-, x^+] = \{x \in \mathbb{R} | x^- \leqslant x \leqslant x^+\}$.

Definition 2 A box $[\mathbf{x}] = ([x_1], \ldots, [x_i], \ldots, [x_n])^{\mathrm{T}} (i = 1, \ldots, n)$ of $\mathbb{IR}^n$ is an interval vector, which is the Cartesian product of n intervals.

Definition 3 The width of $[\mathbf{x}]$ is $w([\mathbf{x}]) = \max_{i=1,\ldots,n} \{x_i^+ - x_i^-\}$.

Definition 4 If $\mathbf{f}$ is a function from $\mathbb{R}^n$ to $\mathbb{R}^m$, the interval function $[f] : \mathbb{IR}^n \to \mathbb{IR}^m$, $[\mathbf{x}] \to [\{f(x) | x \in [\mathbf{x}]\}]$ is an inclusion function of f.

Definition 5 The intersection of the boxes $[\mathbf{x}]$ and $[\mathbf{y}]$ of $\mathbb{IR}^n$ satisfies $[\mathbf{x}] \cap [\mathbf{y}] \triangleq ([x_1] \cap [y_1]) \times \quad \times ([x_n] \cap [y_n])$.

Definition 6 The union of $[\mathbf{x}]$ and $[\mathbf{y}]$ satisfies $[\mathbf{x}] \cup [\mathbf{y}] \triangleq ([x_1] \cup [y_1]) \times \cdots \times ([x_n] \cup [y_n])$.

Definition 7 $[\mathbf{x}] \subset [\mathbf{y}]$ means $[x_1] \subset [y_1]$ and $\ldots$ and $[x_n] \subset [y_n]$.

Definition 8 A subpaving of $[\mathbf{x}]$ is regular if each of its boxes can be obtained from $[\mathbf{x}]$ by a finite succession of bisections and selections.

5.2.2 Set Inversion

Let f be a continuous non-linear function from $\mathbb{R}^n$ to $\mathbb{R}^m$ and $\mathbb{Y}$ be a subpaving of $\mathbb{R}^m$. Set inversion is the operation of calculating the reciprocal image $\mathbb{X}$ of the $\mathbb{Y}$ by function f, expressed as

$$\mathbb{X} = \{x \in \mathbb{R}^n | f(x) \in \mathbb{Y}\} = f^{-1}(\mathbb{Y}). \tag{5.4}$$

SIVIA is the most original set inversion approach to obtain subpavings $\underline{\mathbb{X}}$ and $\overline{\mathbb{X}}$ such that

$$\underline{\mathbb{X}} \subset \mathbb{X} \subset \overline{\mathbb{X}} \tag{5.5}$$

from a sufficiently large initial search box $[\mathbf{x}](0)$.

There are four main steps of SIVIA approach,

1. If $[f]([x]) \subset \mathbb{Y}$, $[x]$ is a feasible box completely belonging to $\mathbb{Y}$, then $[x] \subset \underline{\mathbb{X}}$ and $[x] \subset \overline{\mathbb{X}}$.
2. If $[f]([x]) \cap \mathbb{Y} = \emptyset$, $[x]$ does not belong to $\mathbb{X}$ and is an unfeasible box, then it can be deleted from the solution set.
3. If $[f]([x]) \cap \mathbb{Y} \neq \emptyset$ is established but $[f]([x]) \subset \mathbb{Y}$ is not, $[x]$ is said to be an indeterminate box. If its width is greater than given precision parameters ε, it should be divided into two new boxes as shown in Fig. 5.1 and these newly acquired boxes are applied to a new recursion.
4. If the width of an indeterminate box $[x]$ is less than ε, it is small enough to be stored in $\overline{\mathbb{X}}$.

More details can be found in Fig. 5.1. SIVIA is a recursive algorithm that can get a high precision reciprocal image, but the exponentially increasing calculation time limits its application and development.

5.3 Vector Set Inversion Interval Filter

Construct the state observer of the augmented system (5.3) [6, 7]:

$$\hat{\overline{x}}_k = T\overline{A}\hat{\overline{x}}_{k-1} + T\overline{B}u_{k-1} + L(y_{k-1} - \overline{C}\hat{\overline{x}}_{k-1}) + Ny_k. \tag{5.6}$$

where, $\hat{\overline{x}}_k \in \mathbb{R}^{n+m}$ is the estimated vector of augmented state vector $\overline{x}_k$, $L \in \mathbb{R}^{(n+m)\times q}$ is observer gain, $T \in \mathbb{R}^{n+m}$ and $N \in \mathbb{R}^{(n+m)\times q}$ are constant matrices to be designed, satisfying:

$$T + N\overline{C} = I_{n+m}. \tag{5.7}$$

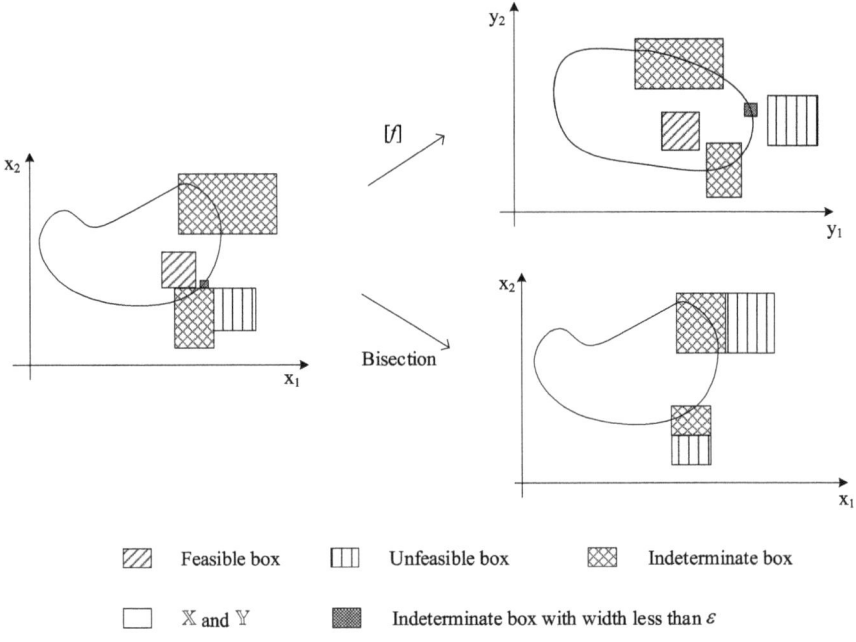

Fig. 5.1 SIVIA diagram

Lemma 2 *The general solution of the matrix T and N in the observer is [6, 7]:*

$$T = M^{\dagger}\Psi_1 + SW\Psi_1, \tag{5.8}$$

$$N = M^{\dagger}\Psi_2 + SW\Psi_2. \tag{5.9}$$

where, $M^{\dagger}$ represents the pseudo-inverse of matrix M, $M = \begin{bmatrix} I_{n+m} \\ \overline{C} \end{bmatrix}$, $\Psi_1 = \begin{bmatrix} I_{n+m} \\ 0 \end{bmatrix}$, $\Psi_2 = \begin{bmatrix} 0 \\ I_q \end{bmatrix}$, $W = I_{n+m+q} - MM^{\dagger}$, $S \in \mathbb{R}^{(n+m)\times(n+m+q)}$ is any matrix.

Define the estimation error as $e_k = \overline{x}_k - \hat{\overline{x}}_k$, the error system can be obtained by substituting the expression (5.7) into the augmented system (5.3):

$$\begin{aligned} e_k &= (T\overline{A} - L\overline{C})e_{k-1} + T\overline{E}w_{k-1} + T\overline{G}\Delta f_{k-1} \\ &\quad - L\overline{D}v_{k-1} - N\overline{D}v_k \\ &= \tilde{A}e_{k-1} + \tilde{B}d_{k-1}. \end{aligned} \tag{5.10}$$

where,

$$\tilde{A} = T\overline{A} - L\overline{C},$$
$$\tilde{B} = \begin{bmatrix} T\overline{E} & T\overline{G} & -L\overline{D} & -N\overline{D} \end{bmatrix},$$
$$d_{k-1} = \begin{bmatrix} w_{k-1}^{\mathrm{T}} & \Delta f_{k-1}^{\mathrm{T}} & v_{k-1}^{\mathrm{T}} & v_k^{\mathrm{T}} \end{bmatrix}^{\mathrm{T}}.$$

To make e_K is robust to disturbance and noise, H_∞ technology is used to design the observer.

Theorem 5.1 *Given a scalar $\gamma > 0$, if there are positive definite matrices $P \in \mathbb{R}^{n+m}$, $Y \in \mathbb{R}^{(n+m)\times n+m+q}$ and $Z \in \mathbb{R}^{(n+m)\times q}$ make*

$$\begin{bmatrix} I_{n+m} - P & * & * & * & * & * \\ 0 & -\gamma^2 I_s & * & * & * & * \\ 0 & 0 & -\gamma^2 I_m & * & * & * \\ 0 & 0 & 0 & -\gamma^2 I_l & * & * \\ 0 & 0 & 0 & 0 & -\gamma^2 I_l & * \\ \Omega_1 & \Omega_2 & \Omega_3 & \Omega_4 & \Omega_5 & -P \end{bmatrix} < 0 \qquad (5.11)$$

hold, the system (5.10) is stable. Where,

$$\Omega_1 = PM^\dagger \Psi_1 \overline{A} + YW\Psi_1 \overline{A} - Z\overline{C},$$
$$\Omega_2 = PM^\dagger \Psi_1 \overline{E} + YW\Psi_1 \overline{E},$$
$$\Omega_3 = PM^\dagger \Psi_1 \overline{G} + YW\Psi_1 \overline{G},$$
$$\Omega_4 = -Z\overline{D},$$
$$\Omega_5 = -PM^\dagger \Psi_2 \overline{D} - YW\Psi_2 \overline{D}.$$

Proof 5.1 Given a scalar $\gamma > 0$, the system (5.10) is stable if and only if there is a positive definite matrix P to for [8]

$$\begin{bmatrix} \tilde{A}P\tilde{A}^T - P + I & * \\ \tilde{B}^T P \tilde{A} & \tilde{B}^T P \tilde{B} - \gamma^2 I \end{bmatrix} < 0 \qquad (5.12)$$

to hold. Using Schur complement lemma and combining $\tilde{A}$ and $\tilde{B}$ in Eqs. (5.10), (5.12) is transformed into

$$\begin{bmatrix} I_{n+m} - P & * & * & * & * & * \\ 0 & -\gamma^2 I_s & * & * & * & * \\ 0 & 0 & -\gamma^2 I_m & * & * & * \\ 0 & 0 & 0 & -\gamma^2 I_l & * & * \\ 0 & 0 & 0 & 0 & -\gamma^2 I_l & * \\ P(T\overline{A} - L\overline{C}) & PT\overline{E} & PT\overline{G} & -PL\overline{D} & -PN\overline{D} & -P \end{bmatrix} < 0 \qquad (5.13)$$

According to the Lemma 2, define

$$\Omega_1 = P(T\overline{A} - L\overline{C}) = PM^\dagger\Psi_1\overline{A} + PSW\Psi_1\overline{A} - PL\overline{C},$$
$$\Omega_2 = PT\overline{E} = PM^\dagger\Psi_1\overline{E} + PSW\Psi_1\overline{E},$$
$$\Omega_3 = PT\overline{G} = PM^\dagger\Psi_1\overline{G} + PSW\Psi_1\overline{G},$$
$$\Omega_4 = -PL\overline{D} = -Z\overline{D},$$
$$\Omega_5 = -PN\overline{D} = -PM^\dagger\Psi_2\overline{D} - PSW\Psi_2\overline{D}.$$

Let $Y = PS$, $Z = PL$, then Eq. (5.11) is established.

For every function $f : \mathbb{R}^a \to \mathbb{R}^b$, there is an interval function $[f] : \mathbb{IR}^a \to \mathbb{IR}^b$, the interval mapping containing the function mapping can be obtained by using interval operations, namely

$$\forall x \in \mathbb{R}^a, [x] \in \mathbb{IR}^a, f(x) \subseteq [f]([x]). \tag{5.14}$$

So according to Eq. (5.10), the interval calculation is performed to calculate the error interval $[e_k]$. Combine the state observer (5.6) to get the observer estimation interval $[\overline{x}_k]^o$ of the augmented state $\overline{x}_k$:

$$[\overline{x}_k]^o = \hat{\overline{x}}_k + [e_k]. \tag{5.15}$$

In the interval calculation process, the iterative calculation using Eq. (5.15) makes the obtained $[\overline{x}_k]^o$ have the wrapping effect. The continuous accumulation of the wrapping effect will cause the final solution to have a great closure. Next, perform interval filtering to further shrink the state interval.

According to the output of the system at time instant k and later at time instant s, the state interval $[\overline{x}_k]^v$ of the system at time instant k is obtained by solving the feasible set X of the following interval set inversion problem:

$$X = \{[\overline{x}_k]^v \in \mathbb{IR}^n | O_{(k:k+s)}[\overline{x}_k]^v \subset [Y_k]\}$$
$$= O_{(k:k+s)}^{-1}[Y_k]. \tag{5.16}$$

where,

$$[Y_k] = y_{(k:k+s)}$$
$$- O_{u(k:k+s)}u_{(k:k+s)} - O_{f(k:k+s)}[\Delta f_{(k:k+s)}]$$
$$- O_{w(k:k+s)}[w_{(k:k+s)}] - O_{v(k:k+s)}[v_{(k:k+s)}],$$

$$O_{(k:k+s)} = \begin{bmatrix} \overline{C} \\ \overline{C}\,\overline{A} \\ \vdots \\ \overline{C}\,\overline{A}^s \end{bmatrix},$$

$$O_{u(k:k+s)} = \begin{bmatrix} 0 & 0 & \cdots & 0 \\ \overline{C}\,\overline{B} & 0 & \cdots & 0 \\ \vdots & \vdots & \vdots & \vdots \\ \overline{C}\,\overline{A}^{s-1}\,\overline{B} & \overline{C}\,\overline{A}^{s-2}\,\overline{B} & \cdots & \overline{C}\,\overline{B} \end{bmatrix},$$

$$O_{f(k:k+s)} = \begin{bmatrix} 0 & 0 & \cdots & 0 \\ \overline{C}\,\overline{G} & 0 & \cdots & 0 \\ \vdots & \vdots & \vdots & \vdots \\ \overline{C}\,\overline{A}^{s-1}\,\overline{G} & \overline{C}\,\overline{A}^{s-2}\,\overline{G} & \cdots & \overline{C}\,\overline{G} \end{bmatrix},$$

$$O_{w(k:k+s)} = \begin{bmatrix} 0 & 0 & \cdots & 0 \\ \overline{C}\,\overline{E} & 0 & \cdots & 0 \\ \vdots & \vdots & \vdots & \vdots \\ \overline{C}\,\overline{A}^{s-1}\,\overline{E} & \overline{C}\,\overline{A}^{s-2}\,\overline{E} & \cdots & \overline{C}\,\overline{E} \end{bmatrix},$$

$$O_{v(k:k+s)} = \begin{bmatrix} \overline{D} & 0 & \cdots & 0 \\ 0 & \overline{D} & \cdots & 0 \\ \vdots & \vdots & \vdots & \vdots \\ 0 & 0 & \cdots & \overline{D} \end{bmatrix},$$

$y_{(k:k+s)}$, $u_{(k:k+s)}$, $[\Delta f_{(k:k+s)}]$, $[w_{(k:k+s)}]$ and $[v_{(k:k+s)}]$ are the output, input, fault difference interval, disturbance interval and noise interval of the system from time instant k to time instant $k + s$, respectively.

As the value of s increases the accuracy of the interval filtering algorithm is higher, but the corresponding calculation amount will continue to increase. To solve this problem, this chapter converts the interval box into the form of a row vector, which can reduce the wrapping effect while ensuring the solution set is unchanged, and reduce the time complexity of the algorithm at the same time. The vector group $\mathcal{L}$ represents all interval boxes in the solution process, and the vector $\mathcal{L}_i (i = 1, 2, 3, \dots)$ represents the i line of $\mathcal{L}$. There are four different situations in the process of using the vector set inversion interval filter to solve Eq. (5.16):

(1) $O_{(k:k+s)}\mathcal{L}_i$ has an intersection with $[Y_k]$ but does not completely belong to $[Y_k]$ and the width of the interval box $[\overline{x}_k]_i^v$ represented by $\mathcal{L}_i$ is greater than the precision parameter ε, the interval box needs to be divided into two new row vectors from the dimension of the maximum interval width.

(2) The intersection of $O_{(k:k+s)}\mathcal{L}_i$ and $[Y_k]$ is empty and the interval box corresponding to $\mathcal{L}_i$ is an infeasible subset.

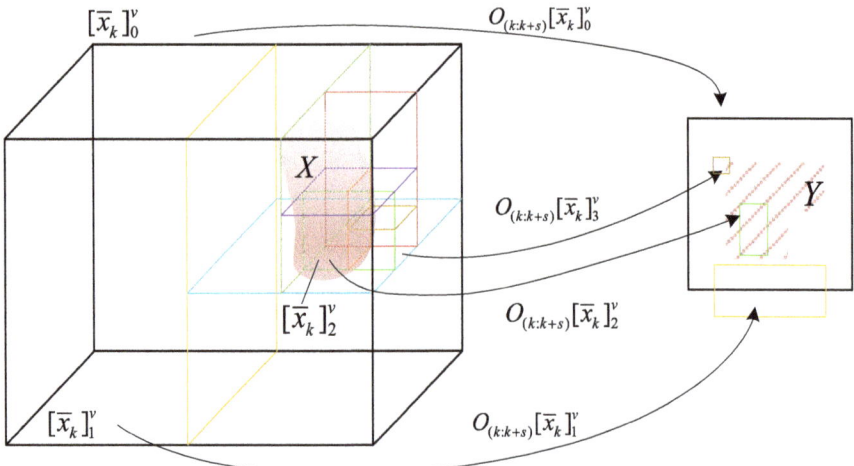

Fig. 5.2 Schematic diagram of vector set inversion interval filtering

(3) $O_{(k:k+s)}\mathcal{L}_i$ completely belongs to $[Y_k]$ and the interval box corresponding to $\mathcal{L}_i$ is a feasible subset.

(4) $O_{(k:k+s)}\mathcal{L}_i$ and $[Y_k]$ have a partial intersection and the width of the corresponding interval box is smaller than the precision parameter ε, the interval box is an uncertain subset.

Figure 5.2 describes the changing trend of interval boxes in the process of vector set inversion interval filtering to solve the feasible set. Where, $[\bar{x}_k]_0^v$, $[\bar{x}_k]_1^v$, $[\bar{x}_k]_2^v$ and $[\bar{x}_k]_3^v$ respectively represent the interval box corresponding to the vector in the above four cases. The irregular body X and the shaded area Y represent the feasible set and the rule set obtained by the feasible set mapping respectively.

For the non-empty vector group $\mathcal{L}$, according to the following test function:

$$[t](\cdot) = \begin{cases} in, & \text{if } O_{(k:k+s)}\mathcal{L}_i \subset [Y_k], \\ out, & \text{if } O_{(k:k+s)}\mathcal{L}_i \cap [Y_k] = \emptyset, \\ eps, & \text{if } W(\mathcal{L}_i) < \varepsilon, \end{cases} \qquad (5.17)$$

where, in, out, eps are all column vectors composed of Boolean variables with the same dimension as $\mathcal{L}$. $W(\mathcal{L})$ is a column vector composed of the width of each interval box corresponding to $\mathcal{L}$ and its dimension is equal to that of $\mathcal{L}$. ε is the given precision parameter. If $O_{(k:k+s)}\mathcal{L}_i \subset [Y_k]$, the boolean variable $in(i) = 1$, otherwise $in(i) = 0$. The interval box corresponding to the vector group $\mathcal{L}(in)$ satisfies the above situation (3), and it is stacked into the feasible set $\mathcal{N}$. In the same way, judge whether $O_{(k:k+s)}\mathcal{L}_i \cap [Y_k] = \emptyset$ holds, and get the Boolean vector $out(i)$. Then the vector group $\mathcal{L}(\neg in \wedge \neg out)$ is an uncertain vector group, denoted by $\mathcal{U}$. If the interval box corresponding to vector $\mathcal{L}_i$ in $\mathcal{U}$ satisfies the above condition (4), the variable $eps(i) = 1$ and stack the interval box corresponding to $\mathcal{U}(eps)$ into the uncertainty layer $\mathcal{E}$. The remaining interval boxes in $\mathcal{U}$ satisfy the above situation (1). After dividing them in two, we get a new vector group $\mathcal{L}$ twice the dimension $\mathcal{U}(\neg eps)$. Loop the entire process until $\mathcal{L}$ is empty, and the vector set inversion interval filtering process is completed. The union of all interval boxes in $\mathcal{N}$ is the filtered state feasible set $[\overline{x}_k]^v$.

In the above process of solving the feasible set X, a series of operations such as interval box stacking, rounding off and bisection are implemented to update the $\mathcal{L}$ search to obtain the solution set through the Boolean operation of the vector group $\mathcal{L}$ without the function recursively after each bisection in the SIVIA algorithm. Assuming that the interval box is divided into two parts at most n times in the process of solving, the computational complexity of the vector set inversion interval filtering algorithm is $O(\log_2(n+1))$, which is much smaller than the computational complexity of SIVIA algorithm $O(2n+1)$, and the computational efficiency is significantly improved.

Theorem 5.2 *The solution set $[\overline{x}_k]^v$ obtained by the vector set inversion interval filtering algorithm satisfies:*

$$[\overline{x}_k]^v \subset X \subset [\overline{x}_k]^v \cup \mathcal{E}. \tag{5.18}$$

Proof 5.2 In the process of solving, if $O_{(k:k+s)}[\overline{x}_k]_i^v$ completely belongs to $[Y_k]$, $[\overline{x}_k]_i^v$ is a feasible subset, which satisfies

$$[\overline{x}_k]_i^v \subset \mathcal{N}, \tag{5.19}$$

The union of all feasible subsets in $\mathcal{N}$ is $[\overline{x}_k]^v$, we can get

$$\bigcup_{i=1,2,\ldots} [\overline{x}_k]_i^v = [\overline{x}_k]^v \subset O_{(k:k+s)}^{-1}[Y_k] = X. \tag{5.20}$$

Similarly, $[\overline{x}_k]_i^v$ is an uncertain subset when $O_{(k:k+s)}[\overline{x}_k]_i^v$ and $[Y_k]$ have a partial intersection and the width of $[\overline{x}_k]_i^v$ is less than the precision parameter ε, it satisfies

$$[\overline{x}_k]_i^v \subset \mathcal{E}, \tag{5.21}$$

All uncertain subsets form an uncertain layer $\mathcal{E}$, satisfying

$$X \setminus [\overline{x}_k]^v \subset \mathcal{E}. \tag{5.22}$$

Therefore,

$$[\overline{x}_k]^v \subset X \subset [\overline{x}_k]^v \cup \mathcal{E}. \tag{5.23}$$

It can be seen from the Theorem 5.2 that the interval $[\overline{x}_k]^v$ based on the vector set inversion interval filtering proposed in this chapter is smaller than the state estimation interval obtained by the commonly used interval filtering algorithm, and the impact of the wrapping effect is lower. According to the above vector set inversion interval filtering method, the state estimation at time instant k is obtained by intersecting the observer estimation interval $[\bar{x}_k]^o$ and the vector set inversion interval filtering contraction interval $[\bar{x}_k]^v$ at that moment:

$$[\bar{x}_k] = [\bar{x}_k]^o \cap [\bar{x}_k]^v. \tag{5.24}$$

Finally, a more compact state interval estimation is achieved.

In summary, the steps of using a fault observer based on vector set inversion interval filtering to diagnose actuator faults are as follows:

Step 1: Construct an augmented system of Eq. (5.3) according to the system model;

Step 2: Design the state observer (5.6), obtain the observer matrix T, L, N by solving the linear matrix inequality (5.10);

Step 3: According to the Eqs. (5.6), (5.10) and (5.15), the interval calculation is performed to obtain the observer estimation interval $[\bar{x}_k]^o$ at the time k;

Step 4: If $k \leqslant N - s$, calculate the rule mapping set Y_k at the time instant k according to the output $y_{(k:k+s)}$, input $u_{(k:k+s)}$, fault difference interval $[\Delta f_{(k:k+s)}]$, disturbance interval $[w_{(k:k+s)}]$ and noise interval $[v_{(k:k+s)}]$ from time instant k to $k + s$:

$$\begin{aligned}
Y_k =& y_{(k:k+s)} \\
& - O_{u(k:k+s)} u_{(k:k+s)} - O_{f(k:k+s)}[\Delta f_{(k:k+s)}] \\
& - O_{w(k:k+s)}[w_{(k:k+s)}] - O_{v(k:k+s)}[v_{(k:k+s)}],
\end{aligned} \tag{5.25}$$

Otherwise, go to Step 6;

Step 5: Use the test function to solve the Eq. (5.16) according to the vector set inversion interval filtering algorithm to get the state interval $[\bar{x}_k]^v$ at the time instant k, and go to Step 7:

Step 6: If $k > N - s$, perform interval calculation to get $[\bar{x}_k]^v$:

$$[\bar{x}_k]^v = \bar{A}[\bar{x}_{k-1}] + \bar{B}u_{k-1} + \bar{E}[w_{k-1}] + \bar{G}[\Delta f_{k-1}]; \tag{5.26}$$

Step 7: According to Eq. (5.24) the state estimation interval $[\bar{x}_k]$ at time instant k is obtained and the actuator fault interval at time instant k is as follows:

$$[f_k] = [0 \quad I_m][\bar{x}_k]; \tag{5.27}$$

Step 8: Set $k = k + 1$, return Step 3; if $k > N$, the algorithm ends.

5.4 Illustrative Simulations

Consider the linear system model as:

$$
\begin{cases}
x_{k+1} = \begin{bmatrix} 0.9842 & 0.0407 \\ 0 & 0.9590 \end{bmatrix} x_k + \begin{bmatrix} 0.0831 & 0.0007 \\ 0 & 0.0352 \end{bmatrix} u_k \\
\qquad + \begin{bmatrix} 0.9842 & 0.0407 \\ 0 & 0.9590 \end{bmatrix} w_k + \begin{bmatrix} 0.8 \\ 0 \end{bmatrix} f_k, \\
y_k = \begin{bmatrix} 0.5 & 0 \\ 0 & 0.5 \end{bmatrix} x_k + \begin{bmatrix} 0.01 & 0 \\ 0 & 0.05 \end{bmatrix} v_k.
\end{cases}
$$

Obtain the observer matrix parameters according to the Theorem 5.1:

$$
T = \begin{bmatrix} 0 & 0 & 0 \\ 0 & 0.6550 & 0 \\ -1.2496 & 0.0086 & 1 \end{bmatrix},
$$

$$
L = \begin{bmatrix} 0 & 0 \\ -0.0393 & 0.3780 \\ -1.2241 & -0.0225 \end{bmatrix},
$$

$$
N = \begin{bmatrix} 2 & 0 \\ 0 & 0.6900 \\ 2.4992 & -0.0173 \end{bmatrix}.
$$

In the simulation, it is assumed that the initial observation state is $\hat{x}_0 = [0 \ \ 0]^T$, the initial error interval is $[e_0] = \begin{bmatrix} [e_{0,1}] \\ [e_{0,2}] \end{bmatrix} = \begin{bmatrix} [-0.1 & 0.1] \\ [-0.1 & 0.1] \end{bmatrix}$, unknown disturbance and noise satisfy $|w_k| \leqslant [0.2 \ \ 0.2]^T$, $|v_k| \leqslant [0.2 \ \ 0.2]^T$, input is $u = [3 \ \ 3]^T$, fault is:

$$
f_k = \begin{cases} 0, & k < 50, k \geqslant 100, \\ 10, & 50 \leqslant k < 100. \end{cases} \tag{5.28}
$$

Figure 5.3 shows the interval estimation result of actuator fault. It can be seen that both the fault observer based on vector set inversion interval filtering and the observer designed in [1] can achieve fault detection and estimation. However, compared to [1], the fault interval obtained by using the vector set inversion interval filtering algorithm proposed in this chapter has a better tracking effect on the fault, and the change of the fault can be detected immediately when the fault occurs and disappears. Moreover the estimation interval of the proposed algorithm is closer to the real fault and less conservative.

Figure 5.4 shows the state estimation data at time instant $k = 38 \sim 45$ in order from bottom left to top right. The dotted rectangle represents the estimated state of the observer, the solid rectangle represents the observer estimation state based on the

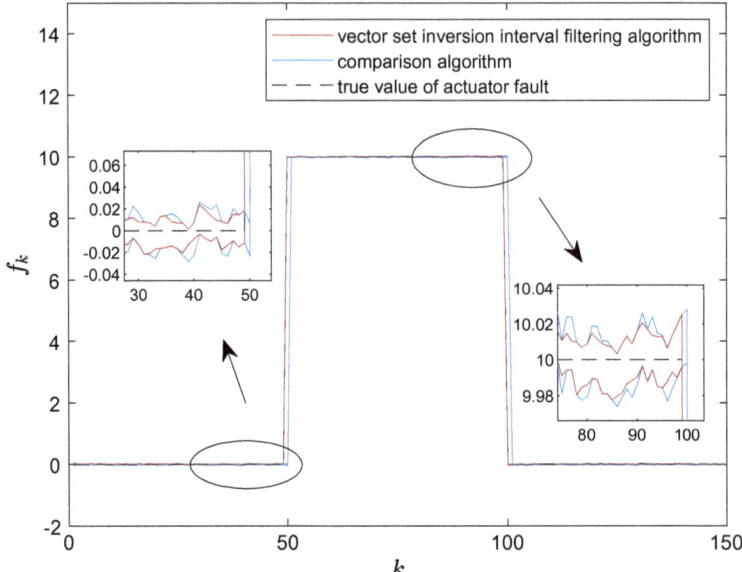

Fig. 5.3 Estimation result of actuator fault interval

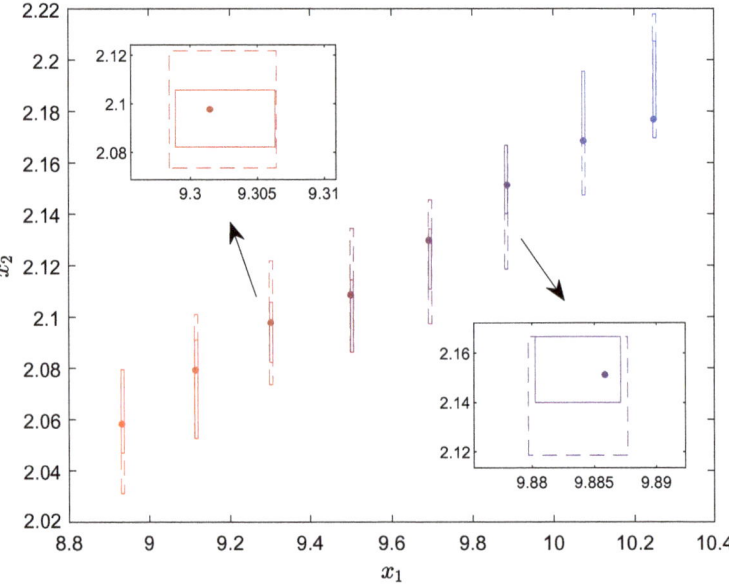

Fig. 5.4 Comparison of state estimation results

vector set inversion interval filtering, and the solid point represents the true value of the current state. It can be seen that the estimated interval after filtering is closer to the true value and the wrapping effect is smaller, indicating that the proposed algorithm can track state changes more effectively.

5.5 Application Study

In order to further verify the feasibility of the actuator fault estimation method, the low-frequency permanent magnet DC motor model is used for simulation analysis, and the structure model is:

$$
\begin{bmatrix} di/dt \\ dn/dt \end{bmatrix} = \begin{bmatrix} -R_a/L & -K_e/L \\ K_t/J_1 & -f_r/J_1 \end{bmatrix} \begin{bmatrix} i \\ n \end{bmatrix} + \begin{bmatrix} 1/L \\ 0 \end{bmatrix} u, \tag{5.29}
$$

where, i and n are motor speed and current respectively, u is the armature voltage. R_a, L, K_e, K_t, f_r and J_1 are the resistance, inductance, back electromotive force constant, torque constant, friction constant and motor inertia of the motor respectively. The specific parameters are shown in Table 5.1.

Take the sampling time $T_s = 1$ ms and discretize the model into the form of system (5.1), the matrix parameters of the system are:

$$
A = \begin{bmatrix} 0.7846 & -0.0154 \\ 0.6056 & 0.9983 \end{bmatrix}, \quad B = \begin{bmatrix} 0.1791 \\ 0 \end{bmatrix}, \quad E = \begin{bmatrix} -0.0085 & -0.0006 \\ -0.0603 & 0.0002 \end{bmatrix}, \quad C = \begin{bmatrix} 1 & 0 \\ 0 & 1 \end{bmatrix},
$$
$$
F = \begin{bmatrix} 0.01 & 0 \\ 0 & 0.01 \end{bmatrix}.
$$

Use the Theorem 5.1 to solve the linear matrix inequality to obtain the observer matrix parameters:

Table 5.1 DC motor model parameters

Variable	Parameter value
R_a	1.2030 Ω
L	5.5840×10^{-3} H
K_e	8.5740×10^{-2} V rad/s
K_t	8.5783×10^{-2} V rad/s
f_r	2.4500×10^{-4} N m s/rad
J_1	1.4166×10^{-4} N m s/rad

$$T = \begin{bmatrix} 0 & 0 & 0 \\ 0 & 0.0262 & 0 \\ -4.9954 & 0.3363 & 1 \end{bmatrix},$$

$$L = \begin{bmatrix} 0.0012 & 0 \\ 0.0080 & 0.0149 \\ -1.8703 & 0.2053 \end{bmatrix},$$

$$N = \begin{bmatrix} 1 & 0 \\ 0 & 0.9738 \\ 4.9954 & -0.3363 \end{bmatrix}.$$

The initial observation state of a given system in the simulation is $\hat{x}_0 = [0 \ \ 0]^T$, the initial error interval is $[e_0] = \begin{bmatrix} [e_{0,1}] \\ [e_{0,2}] \end{bmatrix} = \begin{bmatrix} [-0.06 \ \ 0.06] \\ [-0.6 \ \ 0.6] \end{bmatrix}$, the input is $u_k = 6V$ and the disturbance and noise are $|w_k| \leqslant [0.1 \ \ 0.1]^T \ |v_k| \leqslant [0.1 \ \ 0.1]^T$. Set actuator fault:

$$f_k = \begin{cases} 0, & k < 70, 80 \leqslant k < 160, k \geqslant 200, \\ 0.3u, & 70 \leqslant k < 80, \\ 5, & 160 \leqslant k < 200. \end{cases} \tag{5.30}$$

It can be seen from Fig. 5.5 that vector set inversion interval filtering based fault observer design algorithm can effectively estimate the actuator fault interval of the DC motor. The method in [1] can only detect the fault at the moment after the fault occurs and disappears. In contrast, the proposed method can accurately track the occurrence of two faults and estimate a more accurate fault interval and still has superiority in fault observation accuracy and conservativeness.

Figure 5.6 shows the comparison of the state estimation intervals obtained by sampling every 2 moments in the time instant $k = 140 \sim 150$ from top left to bottom right. It can be seen that the true value of the state represented by the solid point is always within the upper and lower bounds of the estimation interval. At the same time, the dashed rectangular box representing the estimation interval of the observer has always wrapped the real value representing the estimation result of the observer based on the vector set inversion interval filtering algorithm. The line rectangle indicates that the state estimation interval after interval filtering is significantly reduced and closer to the true value.

In order to further verify the effectiveness of vector set inversion interval filtering based fault observer in processing time-varying faults, the initial conditions are kept unchanged and the following actuator faults are set:

$$f_{1,k} = \begin{cases} 0, & k < 70, k \geqslant 200, \\ 2(1 - e^{-0.05(k-70)}), & 70 \leqslant k < 200. \end{cases} \tag{5.31}$$

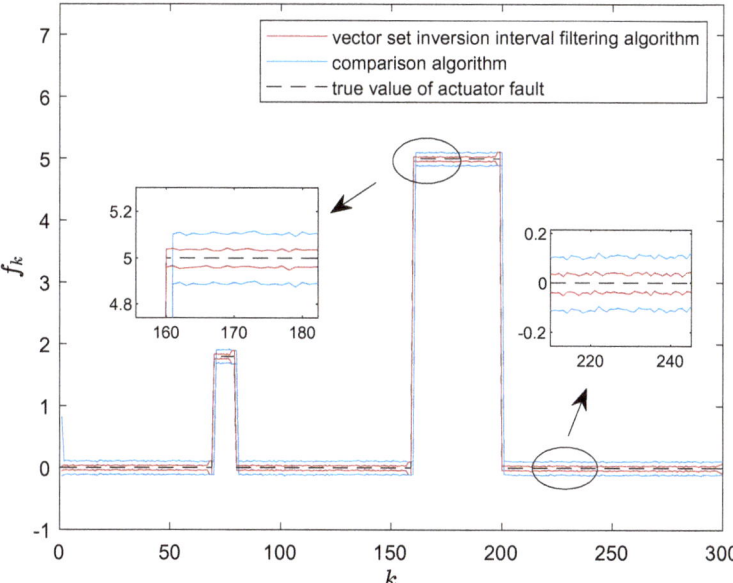

Fig. 5.5 Estimation result of actuator fault interval

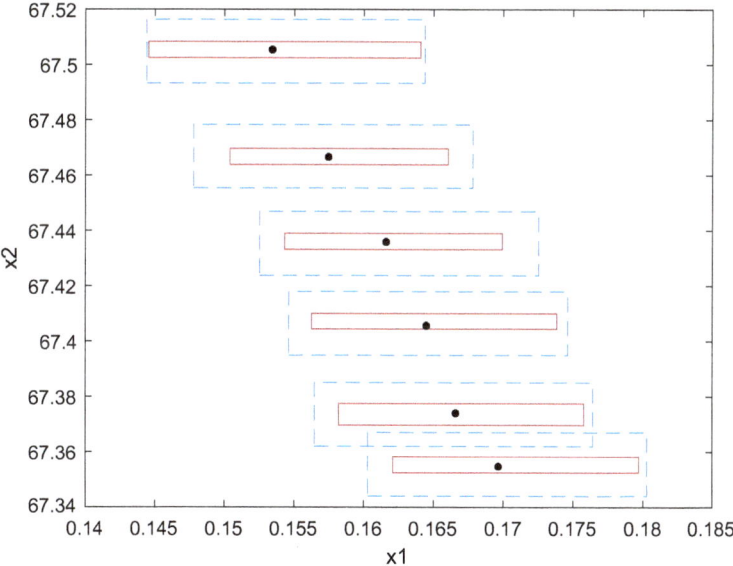

Fig. 5.6 Comparison of state estimation results

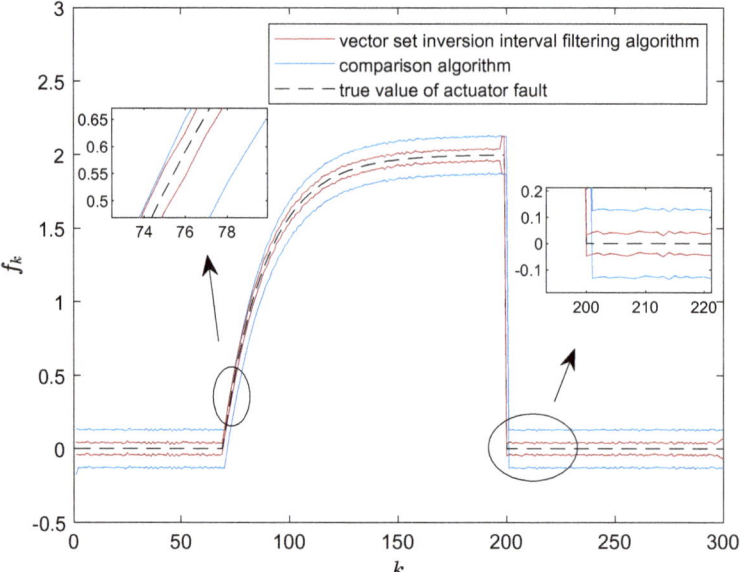

Fig. 5.7 Interval estimation result of actuator fault $f_{1,k}$

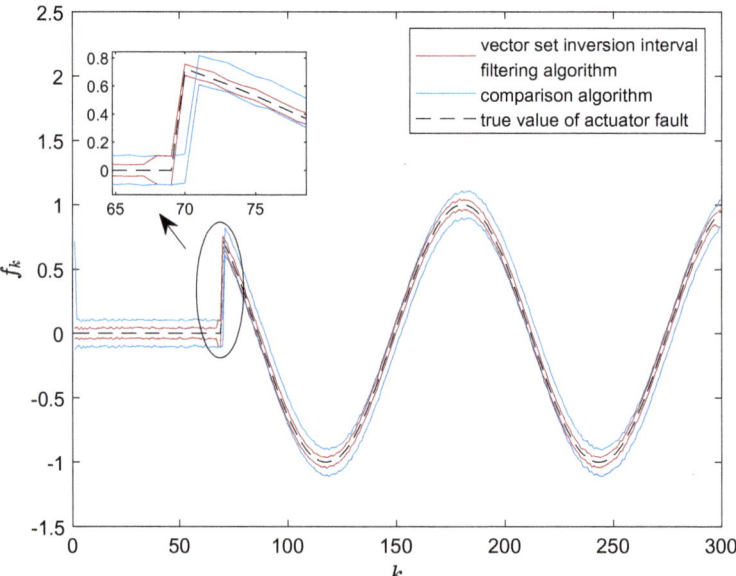

Fig. 5.8 Interval estimation result of actuator fault $f_{2,k}$

$$f_{2,k} = \begin{cases} 0, & k < 70, \\ \sin(0.05(k - 150)), & k \geqslant 70. \end{cases} \tag{5.32}$$

It can be seen from the interval estimation results of actuator exponential fault $f_{1,k}$ in Fig. 5.7 and sine wave type fault $f_{2,k}$ in Fig. 5.8 that the algorithm can not only estimation the abrupt faults effectively, but also track accurately when the fault changes in real time. Due to the function of the vector set inversion interval filtering to shrink the interval, the wrapping effect of the interval observer is reduced and the estimation result is more accurate.

5.6 Concluding Remarks

In this chapter, the observer design and vector set inversion interval filter are combined to study the actuator fault observation method for linear systems with unknown but bounded noise and disturbance. Using the filtering of output data at multiple times to reduce the wrapping effect of interval calculations and solve the problem that the calculation time of traditional interval filtering algorithms increases exponentially with the increase of interval dimensions. Finally, we can intuitively see the feasibility and effectiveness of the method to estimate the actuator fault interval through the simulation of numerical simulation and DC motor system as examples. The fault observer design method proposed in this chapter is suitable for fault estimation problems of other systems with unknown but bounded disturbance and noise, and can also be extended to deal with aircraft systems [9], multi-machine node systems [10], Servo motor system [11], diode circuit [12] and other engineering field fault diagnosis problems.

References

1. Z.H. Wang, M. Rodrigues, D. Theilliol, Actuator fault estimation observer design for discrete-time linear parameter-varying descriptor systems. Int. J. Adapt. Control Signal Process. **29**(2), 242–258 (2015)
2. E. Walter, L. Jaulin, Set inversion via interval analysis for nonlinear bounded-error estimation. Automatica **29**(4), 1053–1064 (1993)
3. R.E. Moore, *Interval Analysis* (Prentice Hall, Englewood Clifs, NJ, USA, 1966)
4. M. Kieffer, E. Walter, Interval analysis for guaranteed nonlinear parameter estimation. IFAC Proc. Vol. **36**(16), 249–260 (1998)
5. Z.P. Qiu, C. Wang, *Interval Analysis of Transient Temperature Field with Uncertain-but-Bounded Parameters*. Mechanics & Astronomy, Science China Physics (2014)
6. Z.H. Wang, C.C. Lim, Y. Shen, Interval observer design for uncertain discrete-time linear systems. Syst. Control Lett. **116**, 41–46 (2018)
7. W.T. Tang, Z.H. Wang, Y. Shen, Interval estimation for discrete-time linear systems: a two-step method. Syst. Control Lett. **123**, 69–74 (2019)

8. W.H. Zhang, Z.H. Wang, Y. Shen, Interval estimation of sensor fault for linear systems. Control Theory Appl. **36**(6), 923–930 (2019)
9. W.H. Zhang, Z.H. Wang, Y. Shen. Interval estimation for sensor fault based on robust positive invariant set. Acta Autom. Sinica (2020)
10. Z.H. Wang, Y. Shen, S.H. Guo, Interval observer design for linear descriptor systems. Control Theory Appl. **35**(7), 956–962 (2018)
11. W.H. Zhang, Z.H. Wang, Y. Shen, Interval estimation of actuator fault by interval analysis. IET Control Theory Appl. **13**(16), 2717–2724 (2019)
12. Y.N. Pan, G.H. Yang, Event-triggered fault detection filter design for nonlinear networked systems. IEEE Trans. Syst. Man Cybern.: Syst. **48**(11), 1851–1862 (2018)

Chapter 6
Design of Orthotopic Set-Membership Based Fault Diagnosis Filter

6.1 Preliminaries and Problem Formulation

Consider the following system:

$$y(k) = \varphi(k)\theta(k) + e(k) = \hat{y}(k) + e(k), \tag{6.1}$$

$$\theta(k) = \theta(k-1) + v(k). \tag{6.2}$$

where, $\theta(k) \in \mathbb{R}^{n_\theta}$, $\varphi(k) \in \mathbb{R}^{1 \times n_\theta}$ are regression vectors that can contain any input $u(k)$ and output $y(k)$, $\theta(k) \in \Theta$ is the parameter vector to be identified, $\theta(k) \in \mathbb{R}^{n_\theta \times 1}$, Θ is the feasible parameter set, $|e(k)| \leqslant \sigma$ is unknown but bounded noise, $\|v(k)\|_\infty \leqslant \gamma$ is the parameter change value.

In this chapter, all feasible parameters that satisfy the model (6.1) are wrapped by the space structure of a zonotope, and the constraint conditions can be expressed as:

$$\Theta_0 = \left\{ \theta \in \mathbb{R}^{n_\theta} \,|\, A_0 \theta \leqslant b_0 \right\}. \tag{6.3}$$

where, $A_0 \in \mathbb{R}^{n \times n_\theta}$, $b_0 \in \mathbb{R}^n$. The formula (6.2) defines the upper bound of the time-varying parameter θ depending on γ. According to the value of γ, the system (6.1) has three classifications:

1. Time-invariant parameters, $\gamma = 0$.
2. Bounded time-varying parameters, $0 < \gamma < \infty$.
3. Unbounded time-varying parameters, $\gamma = \infty$.

In this chapter, we mainly study the problem of parameter identification in bounded time-varying parameter systems, namely $0 < \gamma < \infty$. By constructing the expansion coefficient equation under the constraints of the time-invariant parameter system, the optimal solution γ of the parameter feasible solution set including the parameter change after expansion is used as the final expansion coefficient, instead of simply

© The Author(s), under exclusive license to Springer Nature Singapore Pte Ltd. 2022   75
Z. Wang et al., *Advances in Fault Detection and Diagnosis Using Filtering Analysis*,
https://doi.org/10.1007/978-981-16-5959-1_6

selecting the maximum value of θ as the expansion coefficient, the selection process of the expansion coefficient is optimized. It solves the shortcomings of the high conservativeness of the current time-varying parameters of orthotope identification and the high computational complexity of the time-varying parameters of polyhedron identification.

Consider a IN-CAR system as follow:

$$A(z)y(k) = B(z)f(u(k)) + e(k), \tag{6.4}$$

where $u(k)$ and $y(k)$ are the input and output sequences of the IN-CAR system, respectively; $e(k)$ is an additive noise, which is unknown but assumed to be bounded by a constant $\|e(k)\| \leqslant \delta$; $A(z)$ and $B(z)$ are polynomials in the unit backward shift operator z^{-1} [i.e., $z^{-1}y(k) = y(k-1)$] given by

$$A(z) = 1 + a_1 z^{-1} + a_2 z^{-2} + \cdots + a_{n_a} z^{-n_a},$$
$$B(z) = 1 + b_1 z^{-1} + b_2 z^{-2} + \cdots + b_{n_b} z^{-n_b}.$$

The nonlinear partial equation is

$$\begin{aligned} \bar{u}(k) &= f(u(k)) \\ &= c_1 f_1(u(k)) + c_2 f_2(u(k)) + \cdots + c_{n_c} f_{n_c}(u(k)) \\ &= \sum_{i=1}^{n_c} c_i f_i(u(k)). \end{aligned} \tag{6.5}$$

Suppose the orders n_a, n_b and n_c are known. The purpose of identifying this nonlinear system is proposing an algorithm to consistently estimate the unknown parameter vectors. According to the measured data $\{u(k), y(k)\}_{k=1}^{N}$, the identification model in Eq. (6.4) can be rewritten as

$$\begin{aligned} y(k) &= -\sum_{i=1}^{n_a} a_i y(k-i) + \sum_{i=1}^{n_b} b_i \bar{u}(k-i) + \sum_{i=1}^{n_c} c_i f_i(u(k)) + e(k) \\ &= \varphi^{\mathrm{T}}(k)\theta + e(k), \end{aligned} \tag{6.6}$$

where

$$\begin{aligned} \varphi(k) &= [-y(k-1), \cdots, -y(k-n_a), \bar{u}(k-1), \cdots, \bar{u}(k-n_b), \\ &\quad f_1(u(k)), \cdots, f_{n_c}(u(k))]^{\mathrm{T}} \in \mathbb{R}^{n_a + n_b + n_c}, \\ \theta &= [a_1, \cdots, a_{n_a}, b_1, \cdots, b_{n_b}, c_1, \cdots, c_{n_c}]^{\mathrm{T}} \in \mathbb{R}^{n_a + n_b + n_c}. \end{aligned}$$

From Eq. (6.6), since the noise is bounded, the measurement set at time k can be defined by a strip as follows

$$S(k) = \{\theta : |y(k) - \varphi^{\mathrm{T}}(k)\theta| \leqslant \delta\}, \tag{6.7}$$

which represents a parametric space between two parallel hyperplanes $H_1(k)$ and $H_2(k)$:

$$H_1(k) = \{\theta : \varphi^{\mathrm{T}}(k)\theta = y(k) + \delta\}, \tag{6.8}$$
$$H_2(k) = \{\theta : \varphi^{\mathrm{T}}(k)\theta = y(k) - \delta\}. \tag{6.9}$$

Defined the feasible parameter set at a generic time k as

$$\Theta(k) = \{\theta \in \Theta(0) : |y(k) - \varphi^{\mathrm{T}}(k)\theta| \leqslant \delta, k = 1, \cdots, N\},$$
$$= \bigcap_{t=1}^{k} S(t), \tag{6.10}$$

where $\Theta(0)$ is a feasible parameter set of the orthotopic linear programming (LP) method at the initial moment. Because $\Theta(k)$ is a convex polyhedron in the n-dimensional parameter space, the computation of the minimum orthotope $O(k)$ containing $\Theta(k)$ needs to solve $2n$ LPs with $2k$ constraints.

Based on system (6.6), assume that the fault type is a slowly varying fault of the model parameters. Then $\theta(k)$ is a parameter of the system at time instant k given by

$$\theta(k) = f(\theta_0, k, \alpha), \tag{6.11}$$

where θ_0 is a parameter of the system in the fault-free case. $f(\cdot)$ is a function of θ, k, and α (the variation trend of the fault), when the system's parameter are subject to slowly varying faults. The objective of this chapter is to derive a novel algorithm to detect and isolate the slowly varying faults that occur in IN-CAR systems.

6.2 Orthotope and Its Center

Definition 6.1 Orthotopic approximate feasible parameter set $\mathcal{O}$

$$\mathcal{O}(\bar{\theta}, d) =$$
$$\{\theta : \theta = \bar{\theta} + \mathrm{diag}(d)w, \|w\|_\infty \leqslant 1\}. \tag{6.12}$$

where, $\bar{\theta}, d, w \in \mathbb{R}^n, d_i \geqslant 0, i = 1, \ldots, n_\theta$. $\mathrm{diag}(d)$ is a diagonal matrix whose diagonal value is equal to d, $\mathcal{F}_i = \{\theta \in \mathcal{O} : \theta_i = \bar{\theta}_i + d_i\}$ and $\mathcal{F}_{i+n_\theta} = \{\theta \in \mathcal{O} : \theta_i = \bar{\theta}_i - d_i\}$ is a pair of structural planes of the orthotope $\mathcal{O}$.

It can be seen that the orthotope is determined by the constraint condition θ. By solving $2n_\theta$ linear programming equations, the minimum orthotope $\mathcal{O}^*(\Theta)$ containing the feasible parameter set can be obtained.

$$\beta_i(k) = \max e_i^{\mathrm{T}}\theta,$$

s.t.

$$\theta \in \Theta(k). \tag{6.13}$$

$$\beta_{i+n_\theta}(k) = \min e_i^{\mathrm{T}}\theta,$$

s.t.

$$\theta \in \Theta(k). \tag{6.14}$$

$$\Theta(k) = \bigcap_{i=1}^{2n_\theta} \Theta_i(k). \tag{6.15}$$

where, $\Theta(k)$ is the intersection of $2n_\theta$ constraint solution sets, e_i represents the i column of the n_θ dimensional identity matrix, $i = 1, 2, \cdots, n_\theta$. By solving the linear programming equation, the upper and lower bounds of each parameter in step k can be obtained, and the most compact orthotope $\mathcal{O}^*(\Theta) = \mathcal{O}\left(\overline{\theta}^*, d^*\right)$ can be obtained, where

$$\overline{\theta}_i^* = \frac{\beta_i(k) + \beta_{i+n_\theta}(k)}{2}, \tag{6.16}$$

$$d_i^* = \frac{\beta_i(k) - \beta_{i+n_\theta}(k)}{2}. \tag{6.17}$$

6.3 Linear Programming

This section describes the application of recursive methods to approximate the orthotope $O(k)$ containing $\Theta(k)$ by solving $2n$ linear programming equations at time instant k.

Here some notations are denoted for LP (6.18):

$$\max c^{\mathrm{T}}x,$$

s.t.

$$Ax \leqslant b \tag{6.18}$$

where $\chi = \{x|Ax \leqslant b\}$ denotes the constraint set, $\Xi = \{x \in \chi | x = \arg \max \ c^{\mathsf{T}}x\}$ represents the solution set, a_i^{T} refers to the ith row of matrix A, and $\mathcal{A} : \{x|a_i^{\mathsf{T}} \leqslant b_i, i \in \mathcal{I}\}$ is supposed as the bounding set.

Based on the above definition, some propositions will be used in the following [1].

Proposition 6.1 *Consider the LP*

$$\max \ c^{\mathsf{T}}x,$$
$$\text{s.t.}$$
$$x \in \mathcal{A} \tag{6.19}$$

LPs (6.18) and (6.19) have the same solution and the same solution set Ξ. This means that the constraint set can be replaced by the bounding set without affecting the optimization result.

Proposition 6.2 *If a new constraint* $\mathcal{H} = \{x|a^{\mathsf{T}}x \leqslant b\}$ *is added, then*

$$
\begin{array}{ccc}
\max & c^{\mathsf{T}}x & \geqslant & \max & c^{\mathsf{T}}x. \\
\text{s.t.} & & & \text{s.t.} \\
x \in \mathcal{A} \cap \mathcal{H} & & & x \in \chi \cap \mathcal{H}
\end{array} \tag{6.20}
$$

This shows that, if a new constraint $\mathcal{H}$ is added, a higher degree of conservatism may occur when we replace constraint set χ with bounding set $\mathcal{A}$. The existed work in [1] defines the orthotope $O(k)$, shown by Definition 6.2.

Definition 6.2 Denote the orthotope $O(k)$ as

$$O(\bar{\theta}, d) = \{\theta : \theta = \bar{\theta} + \text{diag}(d)w, \|w\|_\infty \leqslant 1\}, \tag{6.21}$$

where $\bar{\theta}, d, w \in \mathbb{R}^n$, $\text{diag}(d)$ is a diagonal matrix with diagonal equal to d and $d_i \geqslant 0, i = 1, \cdots, n$.

Considering the determination of orthotope $O(k)$ is related to $2k$ constraints, so a compact $O(k)$ is obtain by solving $2n$ LPs:

$$v_i^{(i)}(k) = \max \ e_i^T \theta,$$

$$\text{s.t.}$$

$$\theta \in \mathcal{A}(k) \tag{6.22}$$

$$v_i^{(i+n)}(k) = \min \ e_i^T \theta,$$

$$\text{s.t.}$$

$$\theta \in \mathcal{A}(k) \tag{6.23}$$

$$\mathcal{A}(k) = \bigcap_{i=1}^{2n} \mathcal{A}_i(k), \tag{6.24}$$

where $\mathcal{A}_i(k)$ is the bounding sets of the $2n$ LPs, $v^{(i)}(k)$ is an element belonging to the solution set Ξ, and vector e_i, $i = 1, \cdots, n$, denotes the columns of the identity matrix. Then the orthotope $O(\Theta)$ containing Θ is given by

$$\bar{\theta}_i(k) = \frac{v_i^{(i)}(k) + v_i^{(i+n)}(k)}{2}, \tag{6.25}$$

$$d_i(k) = \frac{v_i^{(i)}(k) - v_i^{(i+n)}(k)}{2}. \tag{6.26}$$

It is assumed that the following are available at time instant k: a set $C(k)$, composed of a subset of the constraints of $\Theta(k)$, i.e., $\Theta(k) \subseteq C(k)$, and a set $C(k) \subseteq O(\bar{\theta}(k), d(k))$. For $2n$ elements, $v^{(i)} \in C(k) \cap \mathcal{F}_i(k)$, where $\mathcal{F}_i(k)$ are the faces of $O(k)$, i.e., $v^{(i)}$ is the maximum estimation value and $v^{(i+n)}$ is the minimum estimation value of the ith parameter for $i = 1, \cdots, n$.

The steps of using linear programming method to achieve orthotope updates at time $k + 1$ are as follows:

Step 1 (LP solution) For each $i = 1, \cdots, 2n$:

(1) If $v^{(i)}(k) \notin S(k+1)$,

$$v^{(i)}(k+1) = \arg \ \max(\min) \ e_i^T \theta,$$

$$\text{s.t.}$$

$$\theta \in C(k) \cap S(k+1) \tag{6.27}$$

where the notation max(min) means that max holds for $i = 1, \cdots, n$, while min holds for $i = n + 1, \cdots, 2n$. $C_i(k+1) = \mathcal{A}_i(k+1)$, and $\mathcal{A}_i(k+1)$ is the bounding set of the LP (6.27).

(2) If $v^{(i)}(k) \in S(k+1)$,

$$v^{(i)}(k+1) = v^{(i)}(k), \tag{6.28}$$

$$C_i(k+1) = C_i(k). \tag{6.29}$$

Step 2 (Orthotope update) Update the orthotope $O(k+1) = O(\bar{\theta}(k+1), d(k+1))$, for $i = 1, \cdots, n$, where

$$\bar{\theta}_i(k+1) = \frac{v_i^{(i)}(k+1) + v_i^{(i+n)}(k+1)}{2}, \qquad (6.30)$$

$$d_i(k+1) = \frac{v_i^{(i)}(k+1) - v_i^{(i+n)}(k+1)}{2}. \qquad (6.31)$$

Step 3 (Set update) Update the set $C(k+1)$:

$$C(k+1) = \bigcap_{i=1}^{2n} C_i(k+1). \qquad (6.32)$$

6.4 Orthotopic Spatial Extension

If $\gamma = 0$, the system is in time invariant state, $\theta(k) = \theta$, at this time, the uncertainty of the system only appears in the unknown but bounded noise. Therefore, for each time k, a hyperplane band F_k containing the parameter space is obtained from the measured value and the regression vector $\varphi(k)$,

$$F_k = \left\{ \theta \in \mathbb{R}^{n_\theta} \mid -\sigma \leqslant y(k) - \varphi(k)\theta \leqslant \sigma \right\}. \qquad (6.33)$$

The orthotope of the k step is obtained by the intersection of all hyperplane bands in $k-1$ steps with the initial orthotope:

$$\mathcal{O}_k^*(\Theta) \supseteq \Theta_k = \left(\bigcap_{i=1}^{k-1} F_i \right) \cap \Theta_0. \qquad (6.34)$$

where, Θ_k is the k step polyhedron, $\mathcal{O}_k^*(\Theta)$ is the orthotope that wraps the polyhedron at step k. Different from solving the constant system parameters, when solving the time-varying parameters, the constraint conditions need to be changed to expand the orthotope, otherwise the orthotope cannot contain feasible solutions after the parameter changes, resulting in identification errors or false alarms as empty sets. In this chapter, the constraint conditions are expanded by the expansion coefficient, and then orthotopes are enlarged and expanded, and finally the goal of expanding the feasible solution set of the parameters is achieved. The expanded orthotopes and polyhedrons are expressed as $\mathcal{O}_k^*(\overline{\Theta})$, $\overline{\Theta}_k$, Define the first $k-1$ step expansion, the k non-expanded orthotope and polyhedron are respectively $\mathcal{O}_k^*(\Theta')$, $\overline{\Theta}'_k$

Definition 6.3 $F_{i|k}(i \leqslant k)$ is the expansion of the hyperplane strip on the basis of the ith moment at the kth moment,

$$F_{i|k} = \{\theta \in \mathbb{R}^{n_\theta} | - (\sigma + \Delta_{i|k}) \leqslant y(i) - \varphi(i)\theta \leqslant \sigma + \Delta_{i|k}\}, \qquad (6.35)$$

where, $\Delta_{i|k}$ is the deviation limit of the output estimate $\Delta(y)_{(i|k)}$ caused by the parameter change from time i to k, and

$$\Delta\hat{y}_{i|k} = \varphi(i)(\theta(k) - \theta(i)) = \sum_{j=i+1}^{k} \varphi(i)v(j). \qquad (6.36)$$

In the time-varying state of the system parameters, considering the parameter change upper bound γ, there is

$$\left|\Delta\hat{y}_{i|k}\right| \leqslant (k - i)\gamma \|\varphi(i)\|_1 = \Delta_{i|k}. \qquad (6.37)$$

Remark 6.1 From Eqs. (6.33) and (6.35), $F_{k|k} = F_k$, that is, the parameter feasible solution set is the same.

The non-expanded orthotope at step k is obtained by the intersection of the extended hyperplane strip at the first $k - 1$ and the initial orthotope:

$$\mathcal{O}_k^*(\overline{\Theta'}) \supseteq \overline{\Theta}'_k = \left(\bigcap_{i=1}^{k-1} F_{i|k}\right) \cap \Theta_0. \qquad (6.38)$$

Corollary 6.1 *Assuming that k is always the last step of identification, it is known that $C(k - 1) = \{\theta : A\theta \leqslant b\}$ and $\mathbf{o}^*(C(k - 1)) = \mathbf{o}(\overline{\theta}, d)$ are polyhedrons and orthotopes on $\mathbb{R}^n$, Then $C(k) \oplus v(k)$ is meaningless, then the feasible set $C(k)$ of parameters at time k satisfies:*

$$C(k) = C(k - 1) \oplus v(k - 1) \subseteq \overline{C} \cap \overline{\mathcal{O}}, \qquad (6.39)$$

where

$$\overline{C} = \{\theta : A\theta \leqslant b + \Delta_{i|k}\}, \qquad (6.40)$$

$$\overline{\mathcal{O}} = \mathbf{o}^*(C) \oplus v(k - 1). \qquad (6.41)$$

Proof 6.1 Assuming that $\theta \in C(k - 1) \oplus v(k - 1)$, there exist $\alpha \in C(k - 1)$ and $\|\beta_i\|_1 \leqslant \|v_i(k - 1)\|_1$, where β_i, $v_i(k - 1)$ represent the i element of β, $v(k - 1)$ respectively, then $\theta = \alpha + \beta$, thereby:

$$A\theta = A(\alpha + \beta) \leqslant b + A\beta \leqslant b + |A|\gamma, \qquad (6.42)$$

When $i = k$, θ only need to meet the k step hyperplane strip, no need to expand; When $i < k$, $|A| < (k - i)\|\varphi(i)\|_1$, where $|A|$ represents the row norm of the matrix A, $\varphi(i)$ represents the i row vector of A, A and b update as the number of steps increases, getting the formula (6.40). So there is $\theta \in \bar{C}$. And because $\alpha \in (C)$, there must be $\alpha \in \mathbf{o}^*((C))$, so

$$\theta = \alpha + \beta \in \mathbf{o}^*(\mathcal{C}) \oplus v(k - 1) = \overline{\mathcal{O}}. \tag{6.43}$$

When $\gamma = 0$, the constraint condition of the orthotope in the step k is expressed as:

$$\Theta_k = \left\{ \theta \in \mathbb{R}^{n_\theta} \middle| \begin{pmatrix} A_0 \\ A_{k-1} \end{pmatrix} \theta \leqslant \begin{pmatrix} b_0 \\ b_{k-1} \end{pmatrix} \right\}, \tag{6.44}$$

$$A_{k-1} = \begin{pmatrix} -\varphi(1) \\ \varphi(1) \\ \vdots \\ -\varphi(k - 1) \\ \varphi(k - 1) \end{pmatrix} \in \mathbb{R}^{2(k-1) \times n_\theta}, \tag{6.45}$$

$$b_{k-1} = \begin{pmatrix} -y(1) + \sigma \\ y(1) + \sigma \\ \vdots \\ -y(k - 1) + \sigma \\ y(k - 1) + \sigma \end{pmatrix} \in \mathbb{R}^{2(k-1)}. \tag{6.46}$$

According to Eq. (6.40) in Corollary 1 and the Eq. (6.37) in Definition 2, when $i = k$, $\Delta_{i|k} = 0$, when $\gamma > 0$, the constraint condition of the non-expanded orthotope in the step k is expressed as:

$$\left\{ \theta \in \mathbb{R}^{n_\theta} \middle| \begin{pmatrix} A_0 \\ A_k \end{pmatrix} \theta \leqslant \begin{pmatrix} b_0 \\ b_k \end{pmatrix} + \gamma \begin{pmatrix} 0 \\ \Delta b_{k-1} \\ 0 \\ 0 \end{pmatrix} \right\}, \tag{6.47}$$

$\Delta b_{k-1} \in \mathbb{R}^{n_{A_k} - 2}$, n_{A_k} is the number of rows of A_k under the condition of extending from 1 to $k - 1$.

The constraint condition of the expanded orthotope in the step k is expressed as:

$$\mathcal{O}_k^*(\overline{\Theta}) \supseteq \overline{\Theta}_k = \left\{ \theta \in \mathbb{R}^{n_\theta} \left| \begin{pmatrix} A_0 \\ A_k \end{pmatrix} \theta \leqslant \begin{pmatrix} b_0 \\ b_k \end{pmatrix} + \gamma \begin{pmatrix} 0 \\ \Delta b_k \end{pmatrix} \right. \right\},$$

(6.48)

where,

$$\Delta b_k = \begin{pmatrix} \Delta b_{k-1;1} + \left\| a_{k;1} \right\|_1 \\ \vdots \\ \Delta b_{k-1;n_{A_k}-2} + \left\| a_{k;n_{A_k}-2} \right\|_1 \\ \left\| a_{k;n_{A_k}-1} \right\|_1 \\ \left\| a_{k;n_{A_k}} \right\|_1 \end{pmatrix},$$

(6.49)

$a_{k;i}$ is ith row of matrix A_k, $\Delta b_{k-1;i}$ is ith value of Δb_{k-1}. At this time, the formula (6.47) can be regarded as a linear programming equation about γ, and the remaining parameters can be obtained from the measured values of the system and the initial orthotope, so γ satisfies the linear programming equation:

$$\gamma_k = \arg(\min_x fx),$$

s.t.

$$\begin{pmatrix} A_0 & 0 \\ & -\Delta b_{k-1} \\ A_k & 0 \\ & 0 \end{pmatrix} x \leqslant \begin{pmatrix} b_0 \\ b_k \end{pmatrix}.$$

(6.50)

where, $f = (0, \cdots, 0, 1)$, $x = \left(\theta^{\mathrm{T}}, \gamma \right)^{\mathrm{T}}$, the length is $n_\theta + 1$. After all γ_k are obtained at the current k time, the final expansion coefficient is $\gamma = \max(\gamma_0, \gamma_k)$.

Figure 6.1 depicts the changing trend of expanded orthotopes and non-expanded orthotope under the same conditions. The dashed-line wrapped area represents the expanded orthotope, the solid-line wrapped area represents the non-expanded orthotope, and the black line wrapped area represents the initial orthotope, because in the case of expanded orthotopes, the constrained hyperplane strip becomes wider. It can be seen that the size of the extended orthotope is larger to include time-varying parameters.

Algorithm steps of the orthotopic spatial extension filtering based system identification algorithm for time-varying parameter systems:

Step1: Define the number of identification steps as L, define the initial orthotope, and get the initial constraints $A_0\theta \leqslant b_0$, $\gamma_0 = 0$;

Step2: According to the input data $u(k)$, the measurement data $y(k)$ and the regression vector $\varphi(k)$ construct the non-expanded orthotopic constraint conditions,

$$A_k = \begin{pmatrix} A_{k-1} \\ -\varphi(k) \\ \varphi(k) \end{pmatrix},$$

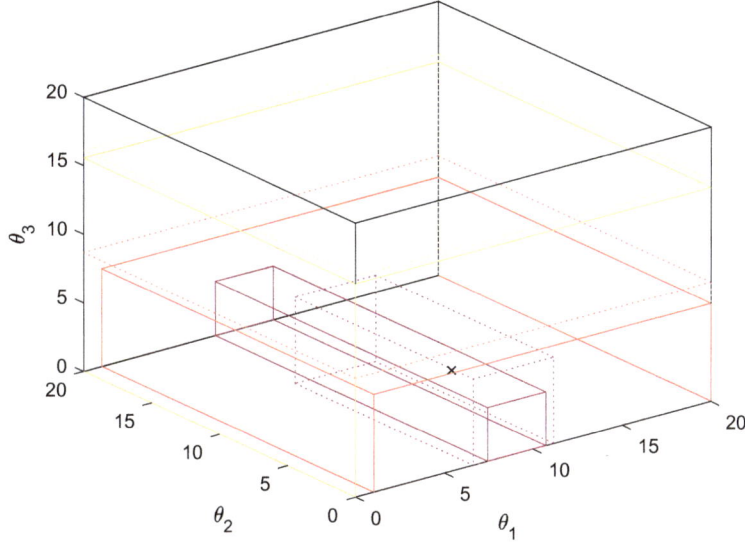

Fig. 6.1 Changes in the spatial expansion of orthotope

$$b_k = \begin{pmatrix} b_{k-1} \\ -y(k) + \sigma \\ y(k) + \sigma \end{pmatrix}, \text{ and } \Delta b_{k-1} = \begin{pmatrix} \Delta b_{k-2} \\ \|\varphi(k-1)\|_1 \\ \|\varphi(k-1)\|_1 \end{pmatrix};$$

Step3: Solve the linear programming Eq. (6.50) on γ, and get the optimal solution γ_k at the step k;

Step4: Update the value of γ, $\gamma = \max(\gamma, \gamma_k)$. Get the maximum expansion coefficient γ when $k = L$, and reset $k = 1$ at the same time, otherwise set $k = k + 1$, return to step 2;

Step5: From Eq. (6.48), update the extended orthotope constraint condition $\overline{\Theta}_k$;

Step6: Respectively take $i = 1, 2, \cdots, n_\theta$ to solve for the vertices of orthotopes,

$$v^{(i)}(k+1) = \arg\max e_i^{\mathrm{T}}\theta,$$

$$\text{s.t.}$$

$$\theta \in \overline{\Theta}_k, \tag{6.51}$$

$$v^{(i+n_\theta)}(k+1) = \arg\min e_i^{\mathrm{T}}\theta;$$

$$\text{s.t.}$$

$$\theta \in \overline{\Theta}_k. \tag{6.52}$$

Step7: Construct extended orthotope $\mathcal{O}_k^*(\Theta) = \mathcal{O}_{k+1}^*(\bar{\theta}(k+1), d(k+1)), i = 1, 2, \cdots, n_\theta$, where

$$\bar{\theta}_i(k+1) = \frac{v_i^{(i)}(k+1) + v_i^{(i+n_\theta)}(k+1)}{2}, \tag{6.53}$$

$$d_i(k+1) = \frac{v_i^{(i)}(k+1) - v_i^{(i+n_\theta)}(k+1)}{2}; \tag{6.54}$$

Step8: Set $k = k + 1$, return to step 5; when $k = L$, the algorithm ends.

6.5 Orthotopic-filtering-based Fault Diagnosis Algorithms

6.5.1 Fault Detection Criterion

In the process of solving for the feasible solution of the LPs, fault detection is realized by detecting whether the feasible set of parameters is empty. In other words, a fault occurs when an inconsistency between the measurement data $S(k_D)$ and the set $O(k_D - 1)$ appears. The fault diagnosis judgement condition is

$$O(k_D - 1) \cap S(k_D) = \emptyset, \tag{6.55}$$

with k_D denoting the time instant of fault detection. In this chapter, it can be seen from Step 1 in Sect. 6.3 that, if any point $v^{(i)}(k_D - 1) \notin S(k_D), i = 1, \cdots, 2n$ or $O(k_D - 1) \cap S(k_D) = \emptyset$, it can be judged that a fault occurs at time k_D.

It is supposed that after $k - 1$ calculations are known for the nonlinear system in Eq. (6.6). Then by specifying the coordinate of its vertices $V_j(k - 1)$, where $j \in \{1, \cdots, 2^n\}$, it is only necessary to determine whether $V_j(k - 1)$ is in $S(k)$ at time k and one only needs to decide whether $V_j(k - 1)$ is between two parallel hyperplanes $H_1(k)$ and $H_2(k)$, given by

$$H_1(k) = \{\theta : \varphi^T(k)\theta = y(k) + \delta\},$$
$$H_2(k) = \{\theta : \varphi^T(k)\theta = y(k) - \delta\}.$$

Let

$$B_{j,1} = y(k) - \varphi^T(k)V_j(k - 1) + \delta, \tag{6.56}$$

and

$$B_{j,2} = -y(k) + \varphi^T(k)V_j(k - 1) + \delta. \tag{6.57}$$

Therefore, we can summarize the following conclusions:

1. If the existence of vertices $V_j(k-1)$ makes $B_{j,1} \geqslant 0$ and $B_{j,2} \geqslant 0$, the system is in a fault-free state.
2. If any point of $V_j(k-1)$ that causes $B_{j,1} < 0$ or $B_{j,2} < 0$, then one can be sure that a fault has occurred.

When a fault is detected, the parameter estimation procedure halts. To then proceed with the fault identification, estimations of the slowly varying parameters is necessary. These can be obtained by enlarging $O(k_D - 1) = O^e(k_D - 1)$ to guarantee $O^e(k_D - 1) \cap S(k_D) \neq \emptyset$.

6.5.2 Fault Isolation and Identification

In this subsection, a fault diagnosis algorithm is proposed by expanding the parameter set and two filtering-based fault diagnosis methods are studied.

The fault diagnosis algorithm based on parameter global expansion filtering (PGE-FFD) mainly depends on expanding the parameter set to a global expansion $O(k_D - 1)$.

The expanded set $O^e(k_D - 1)$ is defined as:

$$O^e(k_D - 1) = \{\theta : v^{e(i+n)}(k_D - 1) \leqslant \theta \leqslant v^{e(i)}(k_D - 1)\},$$

where

$$v^{e(i)}(k_D - 1) = \begin{bmatrix} v_1^{(1)}(k_D - 1) + \beta_1 \\ \vdots \\ v_n^{(n)}(k_D - 1) + \beta_n \end{bmatrix}, \qquad (6.58)$$

$$v^{e(i+n)}(k_D - 1) = \begin{bmatrix} v_1^{(1+n)}(k_D - 1) - \beta_1 \\ \vdots \\ v_n^{(n+n)}(k_D - 1) - \beta_n \end{bmatrix}, \qquad (6.59)$$

and $|\theta(k) - \theta(k-1)| \leqslant \beta_i$, which allows $O^e(k_D - 1)$ to contain the set of parameters at the next moment of fault detection.

The specific change component of the parameter vector when the fault occurs can be determined by the estimation parameter curve. Before this, the time interval for outputting the estimation parameters in the case of a fault occurring is L; that is, after detecting that the system fails at time instant k_d, the subsequent output time instant of estimation parameters is $k_D + L, k_D + 2L, \cdots, k_D + \lambda L, \lambda \in \mathbf{Z}^+$, and the system parameter vector is considered to remain unchanged within the data length of L.

The fault identification can be realized by the PGE-FFD algorithm by judging the trend or slope of the parameter estimation curve. At the same time, it is not necessary to complete the parameter estimation under the fault condition. However, the computational complexity of this method is large. Therefore, an optimized method for expansion of the parameter set is further proposed.

The aforementioned PGE-FFD algorithm has a large amount of calculation, and it takes a long time to detect an existing fault. The following parameter directional expansion filtering fault diagnosis (PDE-FFD) algorithm based on spatial dimension reduction parameter orientation expansion is proposed. In this algorithm, a directional amplification fault diagnosis interval, and the type of gradual failure fault is determined by the slope or the change trend of the fault diagnosis curve.

(1) Fault isolation

We extend the set $O(k_D - 1)$ based on a combination of faults that may occur with all parameters after the fault is detected, getting $O_j^e(k_D - 1)$, $j = 1, \cdots, m$, that is,

$$v^{e(i)}(k_D - 1) = \begin{bmatrix} v_1^{(1)}(k_D - 1) + \gamma_1 \\ \vdots \\ v_n^{(n)}(k_D - 1) + \gamma_n \end{bmatrix}, \tag{6.60}$$

$$v^{e(i+n)}(k_D - 1) = \begin{bmatrix} v_1^{(1+n)}(k_D - 1) - \gamma_1 \\ \vdots \\ v_n^{(n+n)}(k_D - 1) - \gamma_n \end{bmatrix}, \tag{6.61}$$

where $m = n + C_n^2 + \cdots + C_n^{n-1}$ is the sum of the combinations of system param-

eters. $|\theta(k) - \theta(k - 1)| \leqslant \gamma_i$, $\prod\limits_{i=1}^{n} \gamma_i = 0$ and $\sum\limits_{k=k_f}^{N} |\gamma_i(k)| \neq 0$, where k_f denotes the

time instant at which the fault occurred, which means that there is no case where all parameters change at the same time, and the sum of the parameter changes is not zero. Recursively, we can estimate the parameters with the expanded orthotope and a fault can be isolated that corresponds to the orthotope $O_j^e(k_D - 1)$ component of the parameter vector at the earliest time instant k_j for which

$$O_j^e(k_j) \cap O(k_D - 1) \neq \emptyset, \ k_j > k_D - 1. \tag{6.62}$$

Alternatively, the corresponding bounding of the $O_j^e(k_D - 1)$ component of the parameter vector is considered in the fault-free case at the earliest time instant k_j for which

$$O_j^e(k_j) \cap O(k_D - 1) = \emptyset, \ k_j > k_D - 1. \tag{6.63}$$

The fault isolation process is presented in Fig. 6.2. Let us take two-dimensional parameter estimation as an example. Because there is no case in which all the parameters are faulty, there are two combinations of faults; i.e., a fault occurs in either the parameter θ_1 or in the parameter θ_2, so $m = 2$ in this two-dimensional parameter estimation process. Then, respectively, we expand $O(k_D - 1)$ to $O_1^e(k_D - 1)$

Fig. 6.2   The process of fault isolation

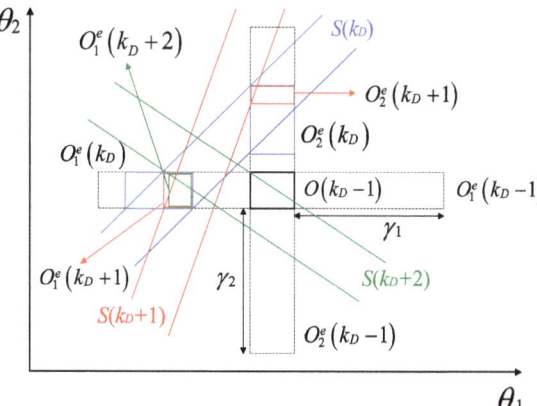

and $O_2^e(k_D - 1)$. It can be seen from Fig. 6.2 that, at the time instant of $k = k_D + 2$, $O_2^e(k_D + 2)$ is empty and $O_1^e(k_D + 2)$ contains parameters. Therefore, fault isolation is completed and the parameter θ_1 is determined to be faulty.

(2) Fault identification
It is supposed that the above method is used to isolate the ith component of the parameter drifts. Then, the parameter set can be extended as

$$v^{e_{(i)}}(k_D - 1) = \begin{bmatrix} v_1^{(1)}(k_D - 1) \\ \vdots \\ v_i^{(i)}(k_D - 1) + \gamma_i \\ \vdots \\ v_n^{(n)}(k_D - 1) \end{bmatrix}, \tag{6.64}$$

$$v^{e_{(i+n)}}(k_D - 1) = \begin{bmatrix} v_1^{(1+n)}(k_D - 1) \\ \vdots \\ v_i^{(i+n)}(k_D - 1) - \gamma_i \\ \vdots \\ v_n^{(n+n)}(k_D - 1) \end{bmatrix}. \tag{6.65}$$

Set the time interval length L for the fault diagnosis process and consider that the system parameter vector has not changed within the length of time L. The fault identification process is realized by judging the parameter estimation curve. If two or more different types of faults occur at the same time, the PDE-FFD algorithm still applies. The PDE-FFD algorithm process can be summarized as Fig. 6.3.

As shown in Fig. 6.3, $O_{j_f}^e(k_D - 1)$ is the directional extended orthotope corresponding to the failed combination in the diagnosis flow chart.

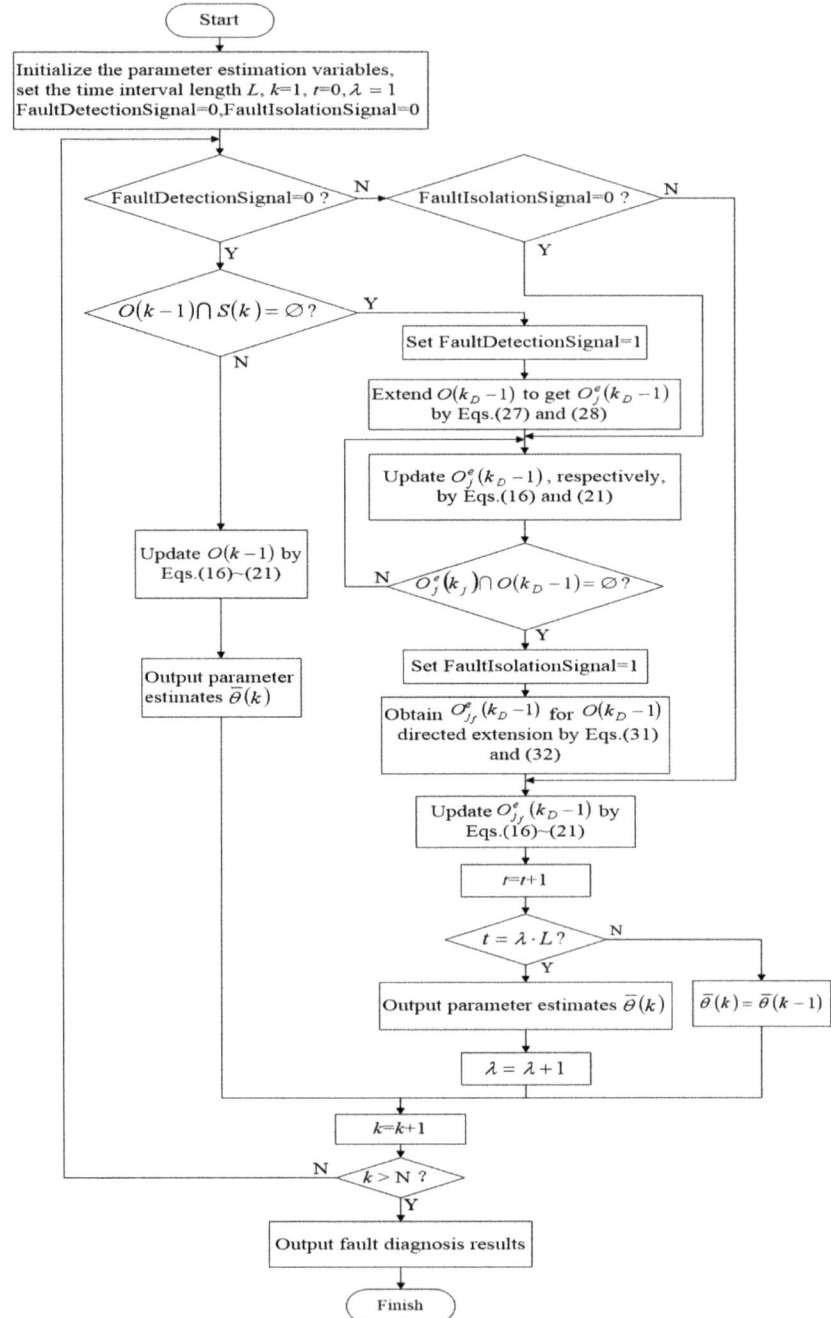

Fig. 6.3 The process of PDE-FFD algorithm

The parameter global expansion filtering fault diagnosis algorithm for expanding all parameters of a faulty system can detect and identify faults. Because the normal parameters are also expanded, the parameter estimation calculation amount in the fault case is the same as the calculation amount in the fault-free case. The proposed parameter directional expansion filtering fault diagnosis algorithm is based on the idea of spatial dimension reduction, making use of a fault isolation filter via directional expansion. After accomplishing fault isolation, only the fault parameters are extended, which reduces the parameter estimation dimension under fault conditions and reduces the amount of calculation needed. Mathematical models and their parameter estimation of dynamical systems are basic for control system analysis and design. Some iterative estimation methods have been proposed [2–5] and can be used to identify the parameters of linear and nonlinear stochastic systems [6–9].

6.6 Hierarchical Fault Diagnosis

6.6.1 Fault Detection

In this chapter, fault free, then the feasible parameter set must not be empty; however, if the feasible parameter set is empty, the system must fail. Therefore, by detecting whether the feasible parameter set is empty allows us to determine whether a fault occurs.

The orthotope of the envelope feasible parameter set is continuously updated and contracted by using the linear programming method. Consider the LPs

$$\max \quad c^{\mathrm{T}}x \tag{6.66}$$
$$\text{s.t.} \quad Ax \leqslant b.$$

The following notation will be used:

(i) $\chi = \{x | Ax \leqslant b\}$ denotes the constraint set.
(ii) $\Xi = \{x \in \chi | x = \arg \max c^{\mathrm{T}}x\}$ represents the solution set of the LP (6.66).
(iii) a_i^{T} refers to the ith row of matrix A.
(iv) $\mathcal{A} : \{x | a_i^{\mathrm{T}} \leqslant b_i, i \in \mathcal{I}\}$ is supposed as the binding set of the LP (6.66).

Definition 6.4 An orthotope $O(t)$ is defined as

$$O(\hat{\theta}, d) = \{\theta : \theta = \hat{\theta} + \operatorname{diag}(d)w, \ ||w||_\infty \leqslant 1\}, \tag{6.67}$$

where $\hat{\theta}, d, w \in \mathbb{R}^n$, $\operatorname{diag}(d)$ is a diagonal matrix with diagonal equal to d, and $d_i \geqslant 0$, $i = 1, \cdots, n$.

In the process of recursive evolution of the feasible parameter set, the linear programming method is used to judge the convergence condition. The method is implemented by solving $2n$ LPs.

Because the orthotope $O(t)$ is determined by $2t$ constraints, the $2n$ LPs need to be solved to obtain the compact orthotope $O(\Theta(t))$ containing $\Theta(t)$ at time t:

$$v_i^{(i)}(t) = \max \quad e_i^T \theta, \tag{6.68}$$
$$\text{s.t.} \quad \theta \in \mathcal{A}(t),$$

$$v_i^{(i+n)}(t) = \min \quad e_i^T \theta, \tag{6.69}$$
$$\text{s.t.} \quad \theta \in \mathcal{A}(t),$$

$$\mathcal{A}(t) = \bigcap_{i=1}^{2n} \mathcal{A}_i(t), \tag{6.70}$$

where $\mathcal{A}_i(t)$ is the binding set of the $2n$ LPs and $v^{(i)}(t)$ is an element belonging to the solution set Ξ. Vector e_i denotes the columns of the identity matrix, $i = 1, \cdots, n$.
Then the orthotope $O(t)$ is given by

$$\hat{\theta}_i(t) = \frac{v_i^{(i)}(t) + v_i^{(i+n)}(t)}{2}, \tag{6.71}$$
$$d_i(t) = \frac{v_i^{(i)}(t) - v_i^{(i+n)}(t)}{2}. \tag{6.72}$$

Based on the above definitions, the following propositions hold for the LPs:

Proposition 6.3 *Consider the LP*

$$\max \quad c^T x \tag{6.73}$$
$$\text{s.t.} \quad x \in \mathcal{A}.$$

LPs (6.66) and (6.73) have the same solution set Ξ.

Proposition 6.4 *If a new constraint $\mathcal{H} = \{x | a^T x \leqslant b\}$ is added, then*

$$\begin{array}{ll} \max \quad c^T x & \geqslant \quad \max \quad c^T x \\ \text{s.t.} \quad x \in \mathcal{A} \cap \mathcal{H} & \quad \text{s.t.} \quad x \in \chi \cap \mathcal{H}. \end{array} \tag{6.74}$$

From Propositions 6.3 and 6.4, we can get that the constraint can be replaced by the bound constraint set without affecting the optimization result. However, if a new constraint $\mathcal{H}$ is added, conservatism may occur. Based on Propositions 1 and 2, we can recursively calculate the orthotope as follows:

Assume that a set $C(t-1)$ is composed of a subset of the constraints of $\Theta(t-1)$, i.e., $\Theta(t-1) \subseteq C(t-1)$ is available at a generic time $t-1$ for $i = 1, \cdots, n$, then there is $C(t - 1) \subseteq O(\hat{\theta}(t - 1), d(t - 1))$. Moreover, for $2n$ elements, $v^{(i)}(t - 1) \in C(t - 1) \cap \mathcal{F}_i(t - 1)$ is also available, that is, $v_i^{(i)}(t - 1) = \hat{\theta}_i(t - 1) + d_i(t - 1)$ and $v_i^{(i+n)}(t - 1) = \hat{\theta}_i(t - 1) - d_i(t - 1)$, where $\mathcal{F}_i(t - 1)$ are the faces of $O(t - 1)$. The above can also be understood as the orthotope $O(t - 1)$ at time $t - 1$ is known, and $O(t)$ at time t is updated by

$$O(t) = O(t - 1) \cap S(t). \tag{6.75}$$

It is necessary to determine whether $O(t)$ is an empty set firstly when the orthotope recursively updates. The system is normal if $O(t) \neq \emptyset$; otherwise, the system fails.

For system (6.6), $v^{(i)}(t - 1)$ are supposed known after $t - 1$ calculations. The the coordinate of its vertices $V_j(t - 1)$ is specified, and $j \in \{1, \cdots, 2^n\}$. Then, it is only necessary to determine whether $V_j(t - 1)$ is in $S(t)$ at time t and we only need to decide whether $V_j(t - 1)$ is between two parallel hyperplanes $H_1(t)$ and $H_2(t)$, as shown in Eqs. (6.8) and (6.9), respectively.

Now let

$$B_{j,1} = y(t) - \varphi^{\mathrm{T}}(t)V_j(t - 1) + \delta, \tag{6.76}$$

$$B_{j,2} = -y(t) + \varphi^{\mathrm{T}}(t)V_j(t - 1) + \delta. \tag{6.77}$$

Case 1 If the existence of $V_j(t - 1)$ makes $B_{j,1} \geqslant 0$ and $B_{j,2} \geqslant 0$, that is, $O(t) \neq \emptyset$, then this indicates that the system is in a fault-free state. The steps to achieve orthotope updates at time t using the linear programming method are as follows:

Step 1 (Solving the LPs): For each $i = 1, \cdots, 2n$,

$$v^{(i)}(t) = \arg \ \max(\min) \ e_i^{\mathrm{T}}\theta \tag{6.78}$$
$$\text{s.t.} \quad \theta \in C(t - 1) \bigcap S(t),$$

where the notation max(min) means that max holds for $i = 1, \cdots, n$, while min holds for $i = n + 1, \cdots, 2n$, $C_i(t) = \mathcal{A}_i(t)$, and $\mathcal{A}_i(t)$ is the binding set of the LP (6.78). Particularly, if $v^{(i)}(t - 1) \in S(t)$ set, then $v^{(i)}(t) = v^{(i)}(t - 1)$ and $C_i(t) = C_i(t - 1)$.

Step 2 (Updating the orthotope): Orthotope $O(t) = O(\hat{\theta}(t), d(t))$ is updated as follows: For $i = 1, \cdots, n$, where

$$\hat{\theta}_i(t) = \frac{v_i^{(i)}(t) + v_i^{(i+n)}(t)}{2}, \tag{6.79}$$

$$d_i(t) = \frac{v_i^{(i)}(t) - v_i^{(i+n)}(t)}{2}. \tag{6.80}$$

Step 3 (Updating the set): Set $C(t)$ is updated as follows:

$$C(t) = \bigcap_{i=1}^{2n} C_i(t). \tag{6.81}$$

Case 2 If any point of $V_j(t-1)$ causes $B_{j,1} < 0$ or $B_{j,2} < 0$, i.e., $O(t) = \emptyset$, then a fault has occurred.

The case of $O(t) = \emptyset$ indicates that all points $v^{(i)}(t-1)$ at time $t-1$ are not in $S(t)$ and so the parameter value has changed greatly, so this determines that a fault has occurred.

Therefore, in the process of solving with LPs, the empty set state of the feasible parameter set is regard as the basis of the fault detection. In other words, a fault occurs when an inconsistency between the measurement data $S(t)$ and the set $O(t-1)$ appears. Therefore, the fault diagnosis logic is

$$O(t-1) \cap S(t) = \emptyset$$

with $t_d = t$ denoting the time instant of fault detection.

The parameter estimation procedure stops when a fault is detected. Proceeding with fault identification requires estimation of the parameters in the fault condition. This can be achieved by reinitializing the system estimation parameters, that is, reinitializing $O(t)$.

Set $t = t + 1$. If a fault is detected at a certain moment, then the fault detection step is terminated, that is, Eqs. (6.78)~(6.81) are directly executed. This is because we consider that the fault will not occur again in the system in a short time. Otherwise, fault detection continues until $t > N$.

The process of applying the linear programming method to update the orthotope and detect the fault is shown in Fig. 6.4.

6.6.2 Fault Identification

In the system with fewer fault samples, the fault-matching method is usually used when performing fault identification. However, for complex systems, a variety of failure types may occur. For a fault library with more fault samples, to reduce the fault diagnosis time, the hierarchical clustering method is first applied to the cluster analysis of the fault types in the fault library.

The condensed clustering algorithm has the advantages of simple calculation, speed, and ease of obtaining similar results. It is not necessary to know the number of clusters in advance, and the algorithm is widely used. Therefore, in this chapter, a condensed clustering algorithm for cluster analysis is used.

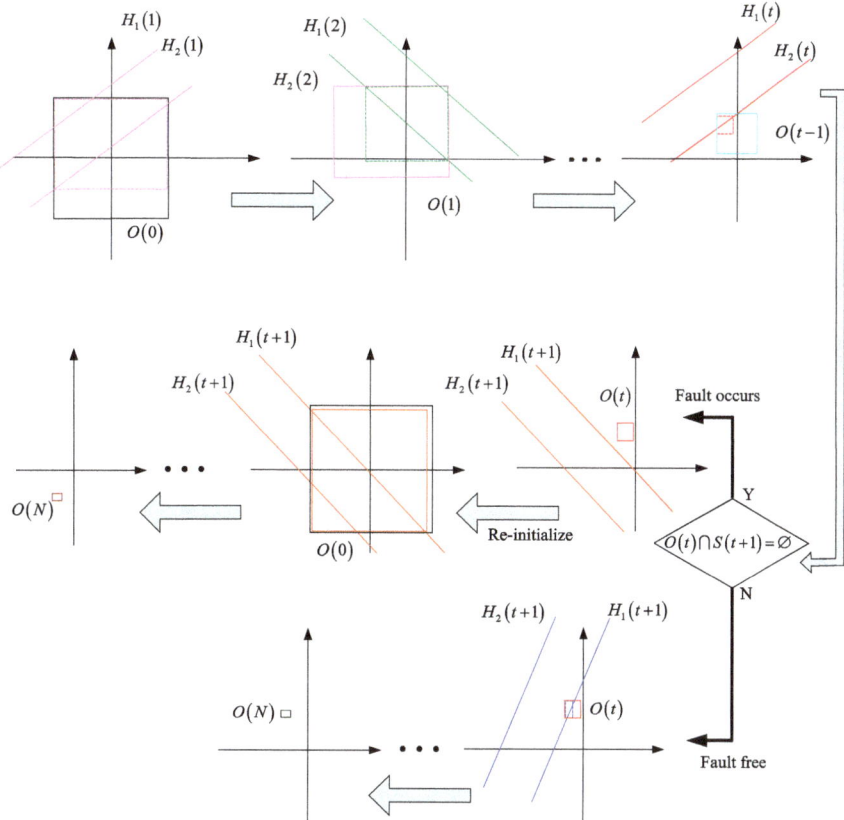

Fig. 6.4 Orthotope updates and fault detection process

Each of the m samples is taken as a class at the beginning of clustering, and the distance between the samples and the cluster between the classes are specified. Then the nearest two classes are merged into a new class, and the new class and other classes are calculated. The distance between the two types is repeated, and one class is reduced each time, until all the samples are merged into one class or the fault library sample cluster is ended when a certain condition is reached.

The class average method is chose to define the distance in this chapter

$$D_{KL}^2 = \frac{1}{n_K n_L} \sum_{x_i \in G_K, x_j \in G_L} d_{ij}^2, \tag{6.82}$$

where G represents a class, with g samples in G, represented by column vector $x_i (i = 1, \cdots, g)$, d_{ij} represents the distance between x_i and x_j, and D_{KL}^2 represents the square distance between class G_K and class G_L.

The recursive formula for the square clustering between classes is given as follow

$$D_{ML}^2 = \frac{n_K}{n_M} D_{KJ}^2 + \frac{n_L}{n_M} D_{LJ}^2. \tag{6.83}$$

The specific process is as follows:

(i) Think of each sample s_i in data set $S = \{s_1, \cdots, s_m\}$ as a cluster class $G_i = s_i$ with a single member, which constitutes a cluster $G = \{G_1, \cdots, G_m\}$ of S.

(ii) Calculate the distance $D(G_i, G_j)$ between each pair of clusters (G_i, G_j) in G to form an initial distance matrix D.

(iii) Select the cluster class pair $\min(D(G_i, G_j))$ with the smallest distance, and merge G_i and G_j into a new cluster class $G_k = G_i \cup G_j$, thus forming a new cluster class $G = \{G_1, \cdots, G_{m-1}\}$ of S and updating the distance matrix.

(iv) Repeat the above steps until there is only one class left in G or the stop condition is met.

Firstly, the following definitions are given before the fault is identified in the fault identification process.

Definition 6.5 Adjacent domain: Two points x and y are neighbors if $sim(x, y) \geq \sigma$, where sim is recorded as a similarity function, σ is the threshold, and sim is selected as the distance metric and can even be selected as a non-metric. The non-metric is normalized so that its value falls between 0 and 1. The larger the value is, the more similar are the two points.

The distance metric to calculate the similarity between categories are chose:

$$sim(x, y) = 1 - \frac{d(x, y)}{d_{max}}, \tag{6.84}$$

where $d(x, y)$ represents the distance between samples x and y and d_{max} denotes the maximum distance between sample x and each sample.

The fault diagnosis will be performed after the fault is detected. The fault diagnosis strategy structure is shown in Fig. 6.5. The steps for applying the hierarchical clustering method for fault identification are as follows:

Step 1: Apply the hierarchical clustering method to the cluster analysis of samples in the fault database.

Step 2: Apply the results of the hierarchical clustering analysis layer by layer. The distance between the estimated parameter $\hat{\theta}(k)$ and the two subclasses of the first layer is calculated by using the average distance method of Eq. (6.82). If the estimated parameter in the continuous data length l period and the distance between one of the subclasses satisfy the condition $D \leq \alpha \cdot \beta$ where α is a threshold index that decreased by a selected constant γ during iterations, and β is hierarchical discriminant initial threshold. The initial value of γ is 0, and 1 is added as the discriminant layer increases by 1, then it is determined that the fault is included in the subclass, and the other subcategory is discarded. The analysis is performed layer by layer in this way until the number of samples in the layer is ≤ 2.

Step 3: Define deviation $\|\hat{\theta}(t - q) - \theta_0\|/\|\theta_0\|$ to satisfy less than a certain threshold within the length of Q to determine the final fault type; that is, $\|\hat{\theta}(t - $

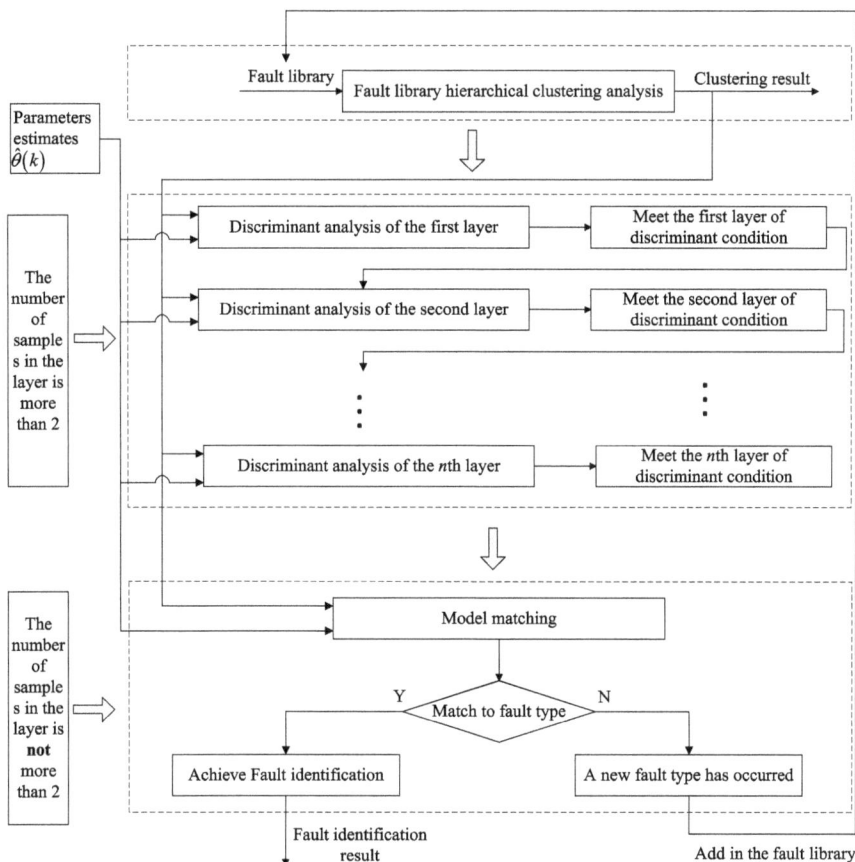

Fig. 6.5 Fault diagnosis strategy structure

$q) - \theta_0||/||\theta_0|| \leqslant \varepsilon(q = 1, \cdots, Q)$ is always established, where θ_0 is a fault sample that matches the last layer and ε is the selected threshold.

Step 4: Assume the simulation length is N. In **Step 3**, if the fault type has not been matched at $N - T_N$ (where T_N is the time threshold left) time instants, then, if the similarity between the other fault type and the matched fault is $\text{sim}(x, y) \geqslant \sigma$ (calculated according to Eq. (6.84)), then the fault type is considered and fault matching is performed at the same time.

Step 5: If the fault of the system is not successfully identified until the end of the simulation, it indicates that a new fault type has appeared in the system at this time. In this condition, this new fault type should be reasonably added to the fault library of the system. When the fault type occurs again, hierarchical cluster analysis is performed on the fault library, and then the fault can be identified according to **Steps 1∼4**.

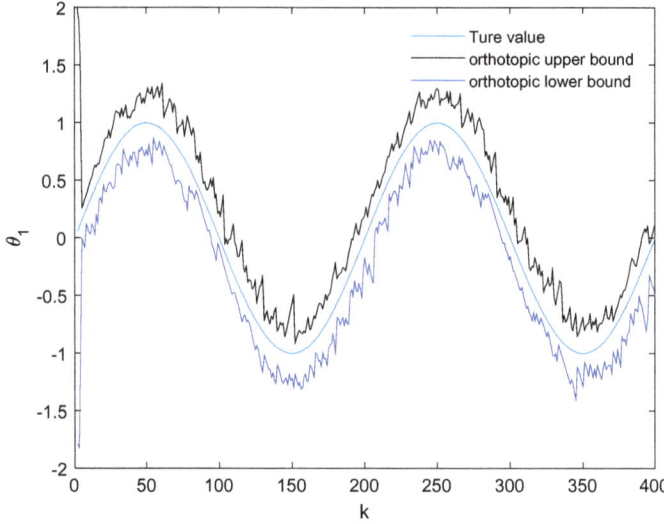

Fig. 6.6 Identification curve of parameter estimate θ_1

6.7   Illustrative Simulations

Example 1   Consider the following linear parameter model:

$$y(k) = \theta_1 u(k) + \theta_2 u(k-1) + e(k), \tag{6.85}$$

where, θ_1, θ_2 are the time-varying parameters to be identified, and the real changes are $\theta_1 = \sin(\pi \times k \times 0.01)$, $\theta_2 = -0.5 \times \sin(\pi \times k \times 0.005)$, $n_\theta = 2$. $e(t)$ is unknown but bounded noise, the input signal $u(t) \in U[-5, 5]$, system noise $e(t) \in U[-0.3, 0.3]$. To compare the changes of tracking parameters, define the Euclidean distance $\Delta\theta = \sqrt{\|\theta - \hat{\theta}\|_1}$, where θ is the true value when the parameter changes, $\hat{\theta}$ is the estimated value of the time-varying parameter [10].

The effect of using the method in this chapter to identify the parameter change value is shown in Fig. 6.6 and Fig. 6.7.

It can be seen from Fig. 6.6 and Fig. 6.7 that the bounded time-varying parameter algorithm based on the orthotopic spatial expansion filter can effectively identify the time-varying parameters. The true value of the parameter is always within the upper and lower bounds of the orthotope. As the parameter changes, the estimated value of the orthotope tracks the change, which has a good effect.

Next set the parameter θ_1 unchanged, the parameter $\theta_2 = -0.5 \times \sin(\pi \times k \times 0.01)$, the simulation results compared with the maximum value of parameter change selected in [1] as the expansion coefficient method and the polyhedron method in [11] are shown in Figs. 6.8~6.9.

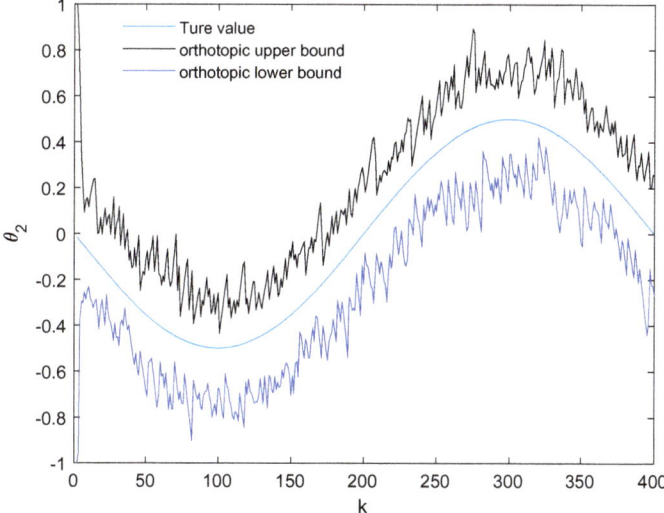

Fig. 6.7 Identification curve of parameter estimate θ_2

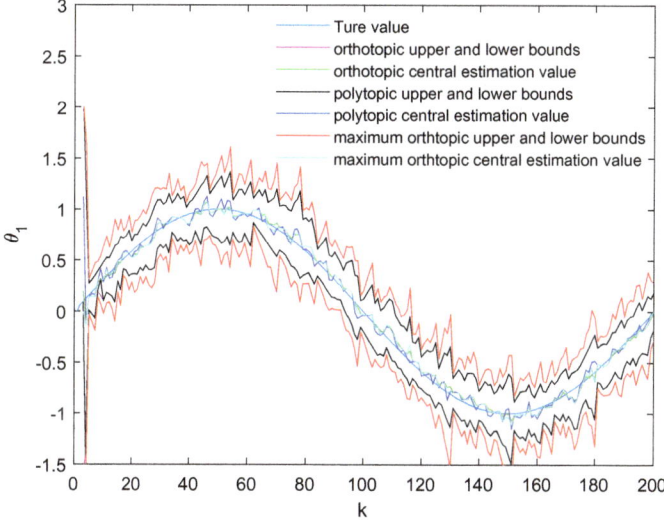

Fig. 6.8 Comparison of time-varying parameter estimate θ_1 between orthotopic and polyhedral based algorithms

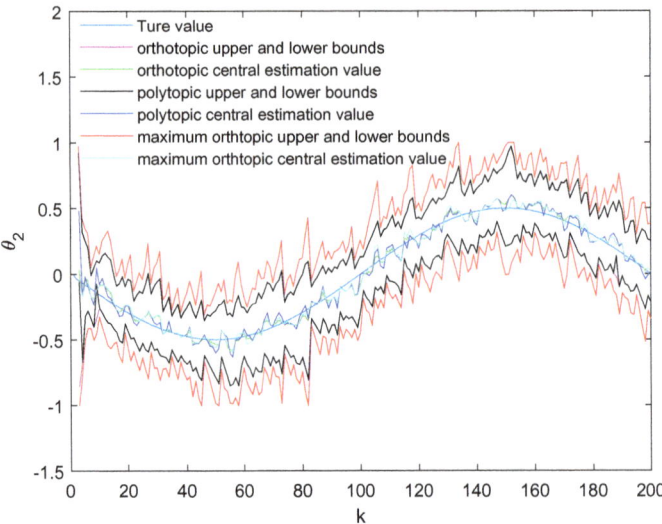

Fig. 6.9 Comparison of time-varying parameter estimate θ_2 between orthotopic and polyhedral based algorithms

From Figs. 6.8 and 6.9, it can be seen that the upper and lower bounds of the feasible parameter solution set of the orthotopic method and the polyhedron method are the same, and the upper and lower bounds of the maximum value of the orthotope are significantly larger than the other two methods. And the central value of the orthotope is more similar to the central value of the face and the maximum parameter change method is closer to the real parameter point. The comparative analysis simulation results compared with the two methods are shown in Figs. 6.10 and 6.11, respectively.

In Fig. 6.10, this chapter takes the data of orthotopes and polyhedrons at intervals of 15 within the range of $k = 50 \sim 140$. The symbol x represents the true value of the parameter at the current moment, and the symbol o represents the estimated value of the center of the orthotope at the current moment, and the symbol * indicates the estimated value of the center of the polyhedron at the current moment, and the symbol □ indicates the estimated value of the center of the maximum orthotope. In Fig. 6.10, the solid line of the standard rectangular wrapping area in different colors indicates the orthotope, the dotted line indicates the maximum orthotope, and the graphic wrapping area within the rectangle indicates the polyhedron. It can be seen that the maximum and minimum values of each parameter in the orthotope are determined according to the upper and lower bounds of the solution set of feasible parameters. Therefore, the spatial expansion filtering method based on the orthotope and the spatial expansion filtering method based on the polyhedron have the same upper and lower bounds when identifying time-varying parameters. In terms of identification accuracy, the estimated value of the center of the orthotope in Fig. 6.10 is closer to the true value of the parameter, reflecting that the method based on the orthotopic

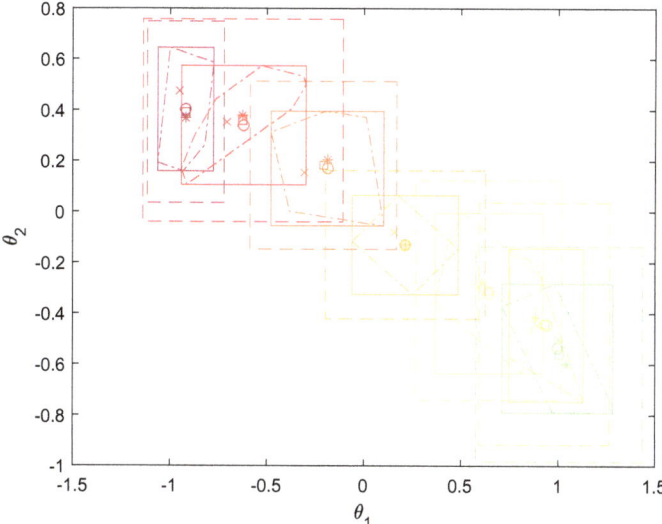

Fig. 6.10   Recursive evolution by the orthotopic and polyhedral algorithms

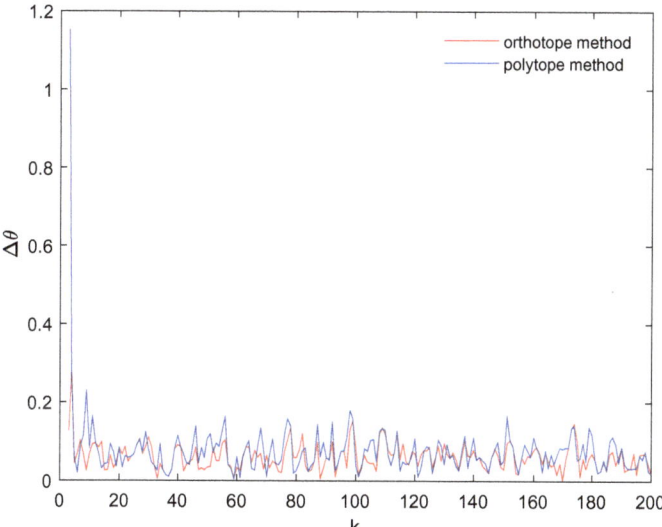

Fig. 6.11   Comparison of error curves by the orthotopic and polyhedral algorithms

Table 6.1 Parameter estimates and errors ($\delta = 0.01$, 0.05 and 0.10)

δ	k	a	b	$\Delta\theta$ (%)
0.01	100	0.58341	0.36154	0.08904
	200	0.58295	0.36197	0.00821
	300	0.58287	0.36202	0.01912
	400	0.58292	0.36202	0.01194
0.05	100	0.58596	0.35966	0.55013
	200	0.58270	0.36181	0.05132
	300	0.58236	0.36214	0.09544
	400	0.58260	0.36214	0.06192
0.10	100	0.59124	0.35723	1.38759
	200	0.58237	0.36159	0.10928
	300	0.58174	0.36235	0.19100
	400	0.58218	0.36235	0.12969
True values		0.58300	0.36200	

spatial expansion filter can achieve more effective tracking of time-varying parameter changes. It can be seen from Fig. 6.11 that in the process of judging the pros and cons of the algorithm based on Euclidean distance, considering that the parameters in the time-varying parameter system change every moment, the estimated value will also fluctuate with the change of the parameter. Therefore, in the error analysis curve, the error curve does not gradually decrease and eventually tends to remain unchanged. However, in the process of error analysis, the estimation error based on the orthotopic spatial expansion filtering method is smaller than that based on the polyhedron spatial expansion filtering method.

Example 2 Consider the following IN-CAR system:

$$A(z)y(k) = \bar{u}(k) + e(k),$$

where $A(z)=1 + az^{-1} = 1 + 0.583z^{-1}, \bar{u}(k) = bf_1(u(k)) = bu^2(k) = 0.362u^2(k)$, $\varphi(k) = [-y(k-1), f_1(u(k))]^{\mathrm{T}}, \theta = [a, b]^{\mathrm{T}} = [0.583, 0.362]^{\mathrm{T}}$.

In the simulation, $u(k) \in N(0, 1)$ and $e(k) \in U[-\delta, \delta]$. We apply the PDE-FFD method proposed in this chapter to estimate the parameters of this nonlinear system. Define the error $\Delta\theta(\%) = \|\bar{\theta}(k) - \theta\|/\|\theta\| \times 100\%$, where θ is the real parameter of the system and $\bar{\theta}$ is the estimated value.

Case 1: Fault-Free Simulation

In the fault free state, the FGE-FFD and PDE-FFD algorithms proposed in this chapter are used to estimate the parameters of the system with $\delta= 0.01$, 0.05, and 0.10, respectively. As we can see, the two algorithms perform in quite the same way for the non fault case set. The estimation results are given in Table 6.1 and Figs. 6.12 and 6.13.

From Table 6.1 and Figs. 6.12 and 6.13, we can get that:

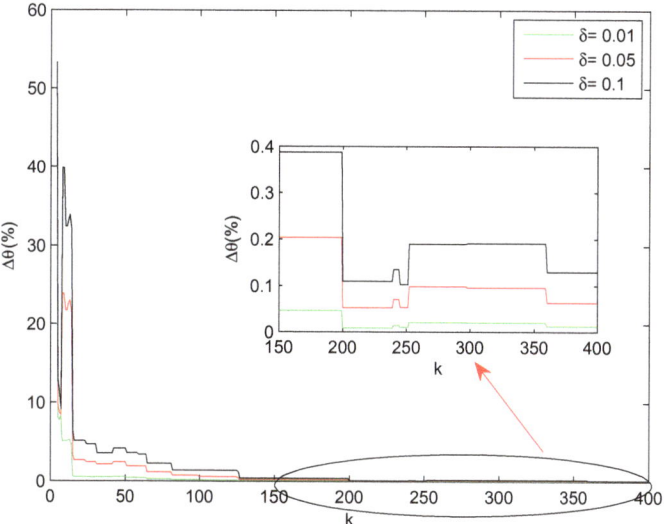

Fig. 6.12 Estimation errors $\Delta\theta(\%)$ versus k with different δ values for the fault-free case

(i) The estimated error is rapidly reduced, which indicates that the estimated parameters converge quickly.

(ii) In the range of $k \in [150, 200]$ and $k \in [300, 350]$, etc., the error is approximately constant; this means that, during this time, every $v^{(i)}(k)$ belongs to set $S(k + 1)$ and the orthotope O does not need to be updated.

(iii) As the data length k increases, the identification accuracy gets higher and the vibration of the estimation error curve becomes gentler and smaller. This indicates that the algorithm can estimate parameters effectively.

(iv) As the noise boundary δ decreases, the proposed method offers better performance and faster convergence.

(v) The recursive evolution process of parameter estimation can be seen more intuitively from Fig. 6.13. For two-dimensional parameters, the orthotope O approximates a two-dimensional box, that is, a square on a plane, and the area is the feasible parameter set in the current estimation step. As the number of recursions increases, O gradually shrinks and eventually converges to a very small box, with its center value being the final parameter estimates.

Case 2: Fault Status Simulation

Based on the model of the fault-free case with $\delta = 0.01$, three typical types of fault are added to parameter b at time instant $k = 200$, respectively. The three faults are fault type 1 (F_1): $b = 0.362 + 0.000006(k - 200)^2$, fault type 2 ($F_2$): $b = 0.362 + 0.002(k - 200)$, and fault type 3 ($F_3$): $b = 0.362 + 0.65 \sin((k - 200)\pi/400)$.

Since the rate of change of the parameters is very small, in order to see the changes of the parameters more clearly, the output interval of the estimation parameters is set as $L = 5$ and the output interval of the selected orthotope is set as $L_1 = 20$.

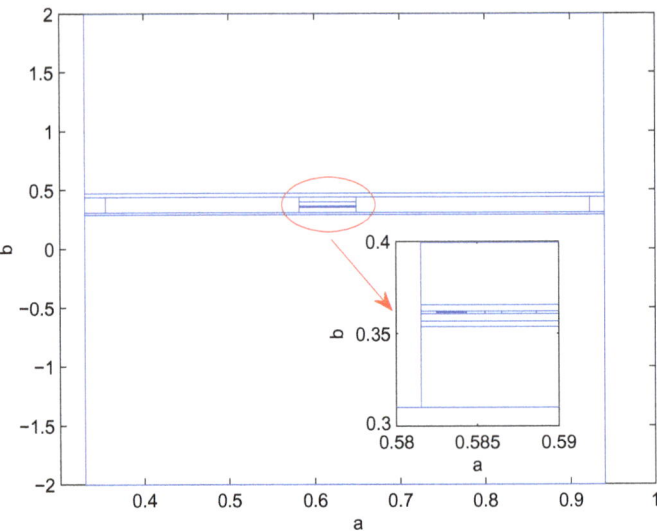

Fig. 6.13 Estimation recursive evolution process of parameters versus k ($\delta = 0.01$) for the fault-free case

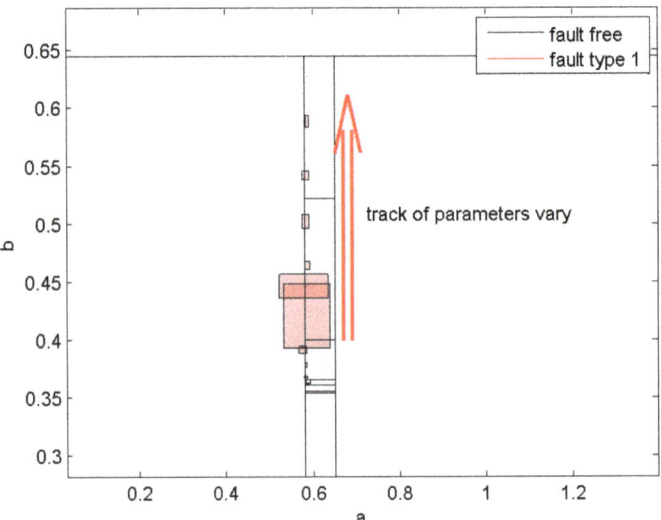

Fig. 6.14 Estimation recursive evolution process of parameters of F_1 versus k obtained by applying the PGE-FFD algorithm

(1) Fault Diagnosis Simulation Results Using the PGE-FFD Algorithm
Applying the PGE-FFD algorithm and taking $\beta_1 = \beta_2 = 0.05$ gives the diagnosis results shown in Figs. 6.14 ∼ 6.18.

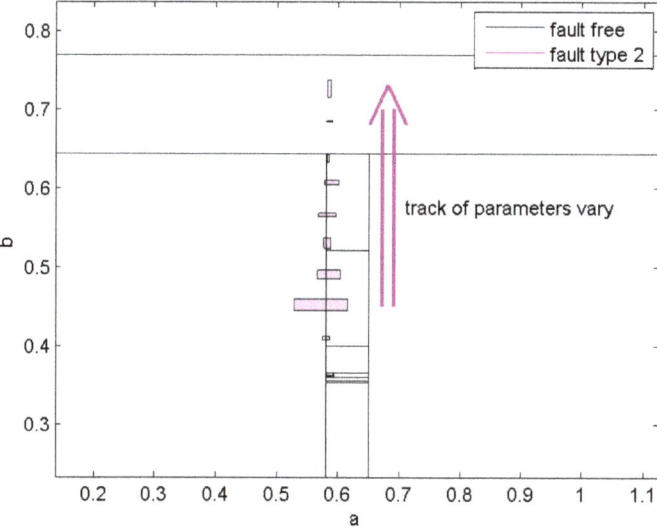

Fig. 6.15 Estimation recursive evolution process of parameters of F_2 versus k obtained by applying the PGE-FFD algorithm

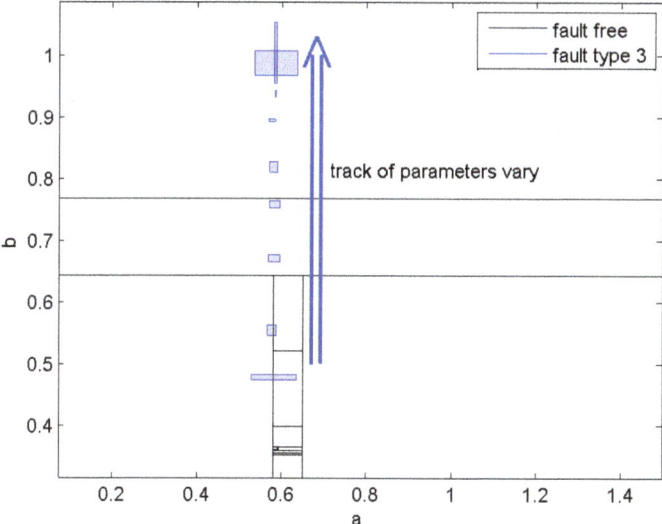

Fig. 6.16 Estimation recursive evolution process of parameters of F_3 versus k obtained by applying the PGE-FFD algorithm

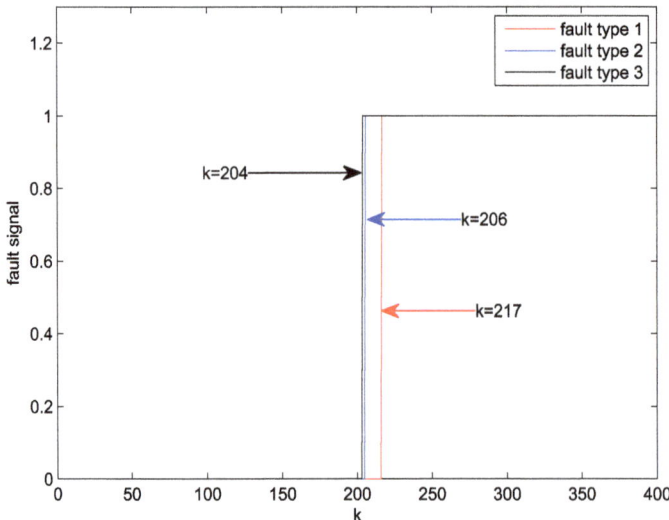

Fig. 6.17 Fault detection signal versus k obtained by applying the PGE-FFD algorithm

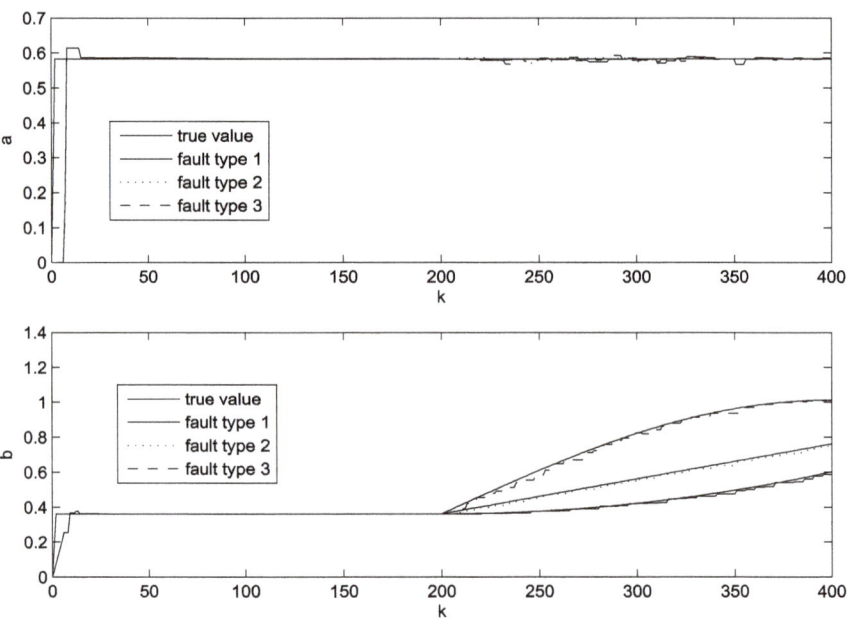

Fig. 6.18 Estimation of parameter tendency versus k obtained by applying the PGE-FFD algorithm

Fig. 6.19 Fault detection and isolation signal versus k obtained by applying the PDE-FFD algorithm

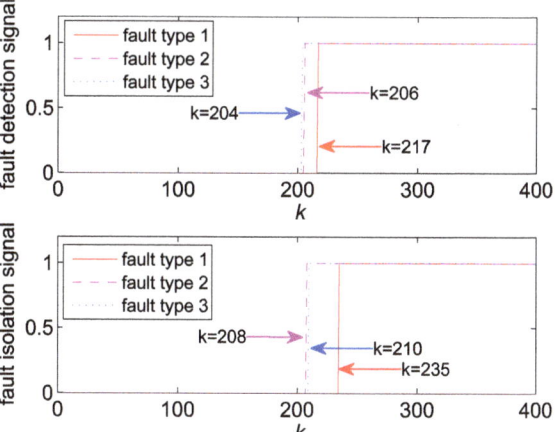

(i) In the fault-free case, it can be seen from Figs. 6.14 ∼ Fig. 6.16, the orthotope O shown as the black boxes gradually shrinks and eventually converges to a very small box as the number of recursions increases, whereas, after the fault occurs, the orthotopes O shown as the red, purple, and blue boxes move at nonuniform speeds in slowly varying directions.

(ii) In the faulty state, the fault detection filter functions and detects a fault at time instants $k_D = 217$, $k_D = 206$ and $k_D = 204$, respectively. For F_1, F_2 and F_3, the fault detection filter jumps as shown in Fig. 6.17. Applying the PGE-FFD algorithm to expand the orthotopes $O(216)$ for F_1, $O(205)$ for F_2 at the same time, and $O(203)$ for F_3. The recursive evolution runs in the direction of the arrow, i.e., red boxes in Fig. 6.14, purple boxes in Fig. 6.15, and blue boxes in Fig. 6.16.

(iii) Fig. 6.18 shows the fault signals and parameter estimates when faults of type $F_1 \sim F_3$ occur individually. It can be seen from Fig. 6.18 that the tendency of parameters estimation is consistent with the trend of the real parameters of the system. Therefore, fault diagnosis can be accomplished by observing the trend of parameter estimation.

(iv) It also can be seen from Fig. 6.18 that, before the fault occurs, after a number of recursive calculations, the parameter estimation value is already close to the true value, but when a fault occurs in parameter b, the estimated value of parameter a also produces a slight fluctuation. This reason for this is that, when the PGE-FFD algorithm is used for fault identification, the estimation interval of the fault-free parameter a is also amplified, which increases the unnecessary calculation amount.

(2) Fault Diagnosis Simulation Results Using the PDE-FFD Algorithm

Applying the PDE-FFD algorithm and taking $\gamma_1 = \gamma_2 = 1$ gives the diagnosis results shown in Figs. 6.19, 6.20, 6.21, 6.22 and 6.23.

(i) From Fig. 6.19, we can get that for different fault types F_1, F_2 and F_3, the fault detection filters detects the fault at time instants $k_D = 217$, $k_D = 206$ and $k_D = 204$, respectively, whereas the fault isolation filters designed according to the method

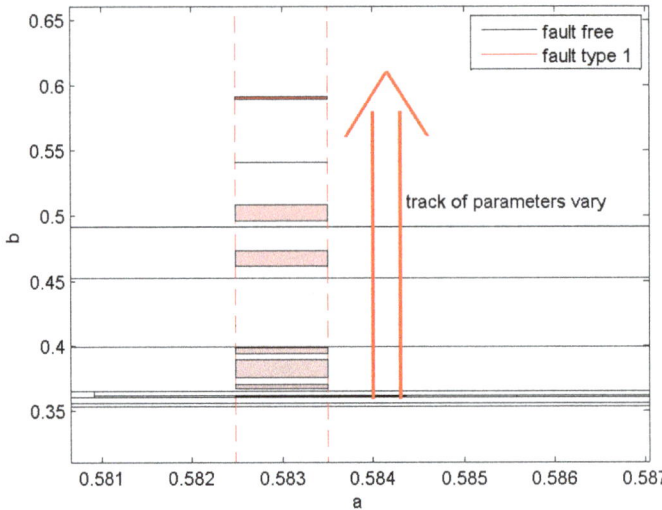

Fig. 6.20 Estimation recursive evolution process of parameters of F_1 versus k obtained by applying the PDE-FFD algorithm

proposed in this chapter achieved fault isolation at time instants $k_I = 235, k_I = 208$ and $k_I = 210$, respectively.

(ii) The estimation recursive evolution process of parameters when F_1, F_2 and F_3 occur separately are shown in Figs. 6.20, 6.21 and 6.22, respectively. After the fault isolation filter isolates when parameter b changes, the parameter directional expansion process is applied to extend the variation range of parameter b in the fault situation, and the estimated values of parameter a remain unchanged, thereby reducing the parameter estimation dimension under the fault condition.

(iii) It can also be seen from Figs. 6.20, 6.21 and 6.22 that the rate of change of F_1 becomes greater faster, the rate of change of F_2 remains unchanged, and the rate of change of F_3 becomes increasingly slower.

(iv) We can know from Fig. 6.23 that the tendency of parameter estimation is consistent with the trend or the slope of the real parameters of the system. Therefore, fault diagnosis can be realized by observing the trend of parameter estimation. In addition, there is no influence on the estimates of the fault-free parameter a during the fault diagnosis process after the fault of parameter b occurs.

In order to further verify the effectiveness and feasibility of the algorithm proposed in this chapter, for a slowly varying fault in the generator and converter dynamic system, the proposed algorithm is used, and compared with the typical recursive least squares (RLS) algorithm to simulation and illustrate.

Consider the generator and converter system of the wind turbine model. The generator and converter dynamic system can be modeled as in as

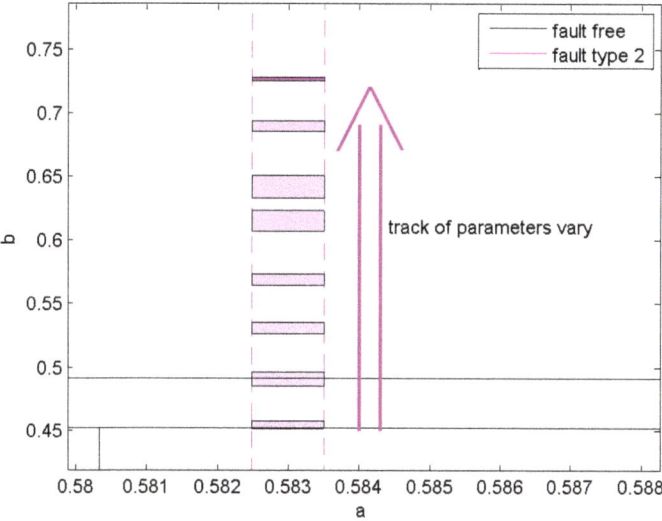

Fig. 6.21   Estimation recursive evolution process of parameters of F_2 versus k obtained by applying the PDE-FFD algorithm

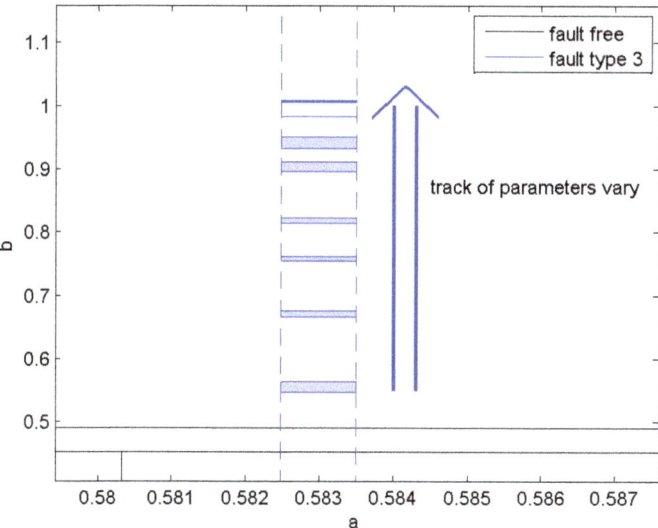

Fig. 6.22   Estimation recursive evolution process of parameters of F_3 versus k obtained by applying the PDE-FFD algorithm

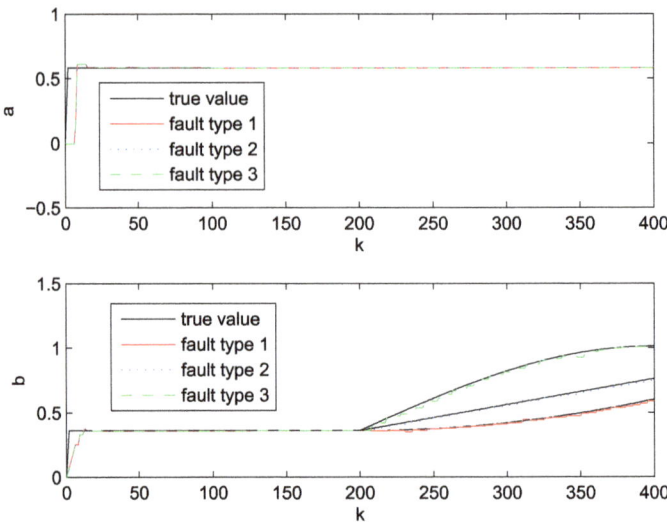

Fig. 6.23 Estimation of parameter tendency versus k obtained by applying the PDE-FFD algorithm

$$\frac{\tau_g(s)}{\tau_{g,r}(s)} = \frac{\alpha_{gc}}{s + \alpha_{gc}}, \tag{6.86}$$

where τ_g and $\tau_{g,r}$ are the generator torque and its reference, respectively, and α_{gc} is the given generator and converter model parameter. We set the parameter to $\alpha_{gc} = 1$ and discrete time to $T_s = 0.01$. Adopting bilinear transformation to discretize Eq. (6.86) yields the equation

$$\frac{\tau_g(k)}{\tau_{g,r}(k)} = \frac{0.005 + 0.005z^{-1}}{1 - 0.990z^{-1}}, \tag{6.87}$$

replacing $\tau_g(k)$ and $\tau_{g,r}(k)$ in Eq. (6.87) with $y(k)$ and $\bar{u}(k)$, respectively, gives

$$\frac{y(k)}{\bar{u}(k)} = \frac{0.005 + 0.005z^{-1}}{1 - 0.990z^{-1}}$$
$$:= \frac{b_0 + b_1 z^{-1}}{1 + az^{-1}}. \tag{6.88}$$

After using the normalization method in [12] and considering the system noise $e(k)$, Eq. (6.88) can be rewritten as

$$A(z)y(k) = B(z)\bar{u}(k) + e(k),$$

where $A(z)=1+az^{-1}=1-0.990z^{-1}$, $B(z)=1+bz^{-1}=1+0.005z^{-1}$, $\bar{u}(k)=$ $cf_1(u(k))=cu^3(k)=0.067u^3(k)$, $\theta=[a,b,c]^{\mathrm{T}}=[-0.990,0.005,0.067]^{\mathrm{T}}$, $\varphi(k)=$ $[-y(k-1),\bar{u}(k-1),f_1(u(k))]^{\mathrm{T}}$.

In the simulation, $u(k)\in \mathrm{N}(0,1)$ and $e(k)\in \mathrm{U}[-0.01,0.01]$. We take $L=5$ and set the output interval of the selected orthotope to $L_1=30$.

Case 3: Single-Fault Status Simulation

Based on the model of the fault-free case, one type of fault is added to parameter a at time instant $k=200$: $a=-0.990+0.5\sin((k-200)\pi/400)$.

(1) Fault Diagnosis Simulation Results Using the PGE-FFD Algorithm

Applying the PGE-FFD algorithm and taking $\beta_1=\beta_2=0.05$ gives the diagnosis results shown in Figs. 6.24 and 6.25.

(i) As shown in Fig. 6.24, we can be seen from the two-dimensional plot that when the system is in a fault-free condition, the black box is gradually reduced to a very small one. When the fault occurs, the fault detection filter detects the fault at time instant $k_D=209$, and then the various estimated parameters are expanded, and the parameter estimation under the fault condition is continued as the red boxes move. It can be seen that the red boxes gradually move in the same direction in which the a axis increases, in accordance with the variation of parameter a.

(ii) From Fig. 6.25, we can get that when the parameter changes slowly, the estimated parameters obtained by the RLS algorithm fluctuate greatly around the true value, the tracking effect is poor, and fault diagnosis cannot be implemented well. On the contrary, the tendency of parameter estimation obtained by the PGE-FFD algorithm is consistent with the variation of the real parameters of the system. Similarly, the estimation of fault-free parameters b and c is affected during the fault diagnosis process, and the estimated values generate slight fluctuations around their true values.

(2) Fault Diagnosis Simulation Results Using the PDE-FFD Algorithm

Applying the PDE-FFD algorithm and taking $\gamma_1=\gamma_2=1$ gives the diagnosis results shown in Figs. 6.26 $\sim$ 6.30.

(i) In Fig. 6.26, $f_1\sim f_3$ are correspond to the fault isolation filters of parameter a, b and c, respectively. The fault detection filter produces a jump at time instant $k_D=209$ in Fig. 6.26, indicating that the fault is detected. The fault isolation filter for parameter a produces a jump at time instant $k_I=211$, representing that fault isolation is accomplished.

(ii) In Fig. 6.27, the estimation recursive evolution process of parameters in the fault-free case (as shown in the black boxes) and fault status (as shown in the red boxes). The parameter estimation process in the fault-free case is the same as that for the PGE-FFD algorithm and will not be described here again. When the isolation process ends and it is detected that parameter a has failed, only parameter a is expanded. During the estimation of the fault state, the estimates of the other parameters b and c remain unchanged, and the estimated value of parameter a varies along the direction in which the a axis increases.

(iii) Figs. 6.28 and 6.29 show the expansion process and parameter estimation process under fault conditions more clearly. The red dotted box in the figure is the

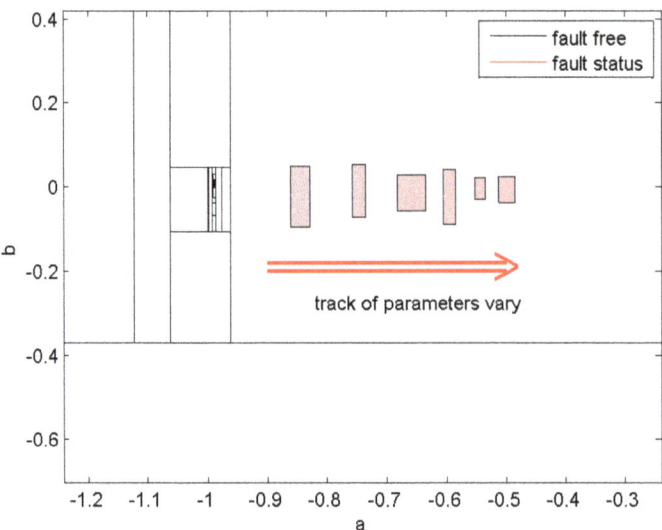

Fig. 6.24 Estimation recursive evolution process of parameters a and b versus k obtained by applying the PGE-FFD algorithm

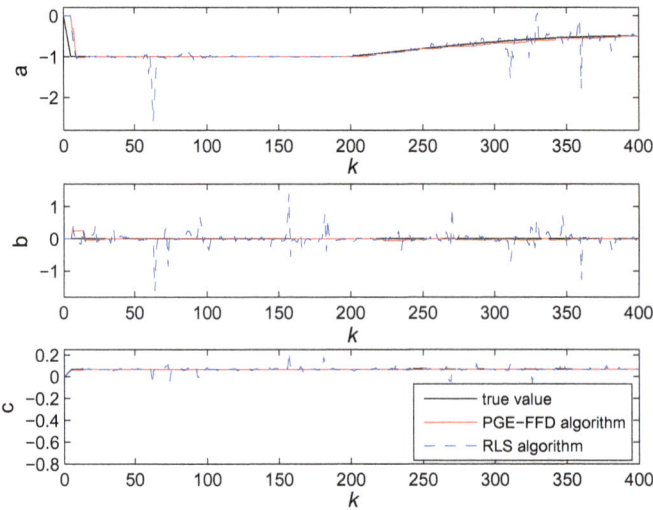

Fig. 6.25 Estimation of parameter tendency versus k obtained by applying the PGE-FFD algorithm

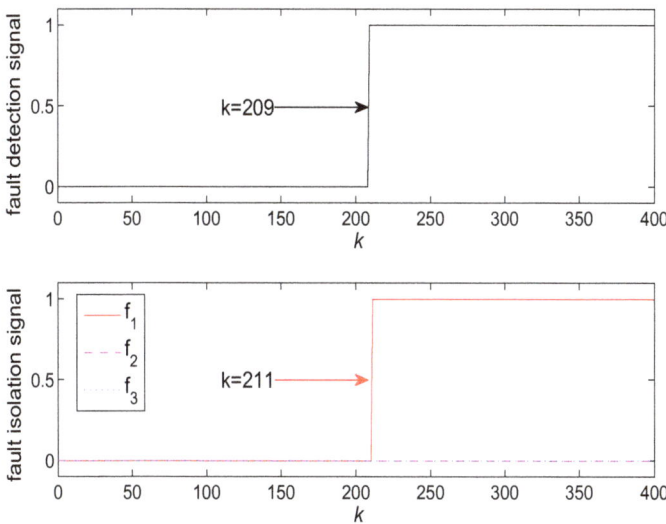

Fig. 6.26 Fault detection and isolation signal versus k obtained by applying the PDE-FFD algorithm

extended range. The red box is a feasible solution for parameter a corresponding to the moment and moves in the direction of the arrow in the box with the red dotted line, thereby reducing the dimension of the estimates.

(iv) It can be seen from Fig. 6.30 that the tendency of parameter estimation is consistent with the trend or the slope of the real parameters of the system. Moreover, it also can be seen that, after a slowly varying fault occurs in parameter a, there is no influence on the estimates of parameters b and c during the fault diagnosis process. Instead, RLS algorithm has poor tracking effect.

Case 4: Multiple-Fault Status Simulation
This chapter does not consider the relatively simple full-fault situation in this simulation. Based on the model of the fault-free case, one type of fault is added to parameters a and c at time instant $k = 200$: $a = -0.990 + 0.5 \sin((k - 200)\pi/400)$ and $c = 0.067 - 0.000025(k - 200)^2$.

The PGE-FFD algorithm for multiple-fault situations is similar to the PGE-FFD algorithm for single faults. Only the PDE-FFD algorithm is applied here for multiple-fault diagnosis. We apply the PDE-FFD method and take $L = 5$ and $\gamma_1 = \gamma_2 = 1$. We set the output interval of the orthotope to $L_1 = 30$. Performing fault diagnosis according to Algorithm 2 then gives the diagnosis results shown in Figs. 6.31, 6.32, 6.33, 6.34 and 6.35.

(i) $f_1 \sim f_3$ in Fig. 6.31 correspond to the fault isolation filter of parameters $\{a, b\}$, $\{b, c\}$ and $\{a, c\}$, respectively. It can be seen from Fig. 6.31 that the fault detection filter produces a jump at time instant $k_D = 209$, indicating that the fault is detected. The fault isolation filter f_3 for parameter $\{a, c\}$ gives rise to a jump at time instant $k_I = 216$, representing that fault isolation is achieved.

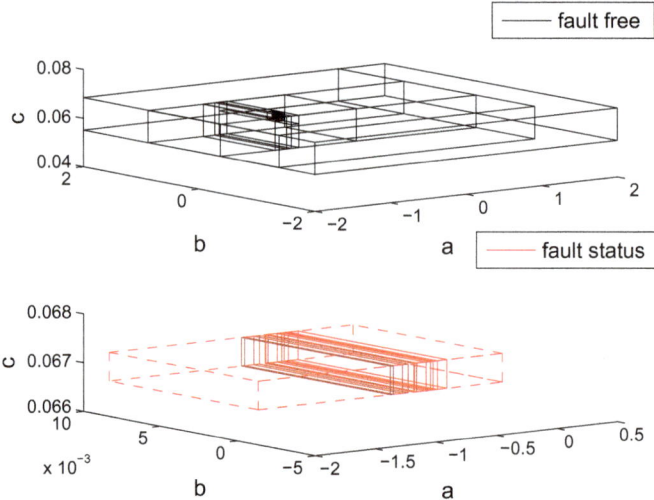

Fig. 6.27 Estimation recursive evolution process of parameters versus k obtained by applying the PDE-FFD algorithm

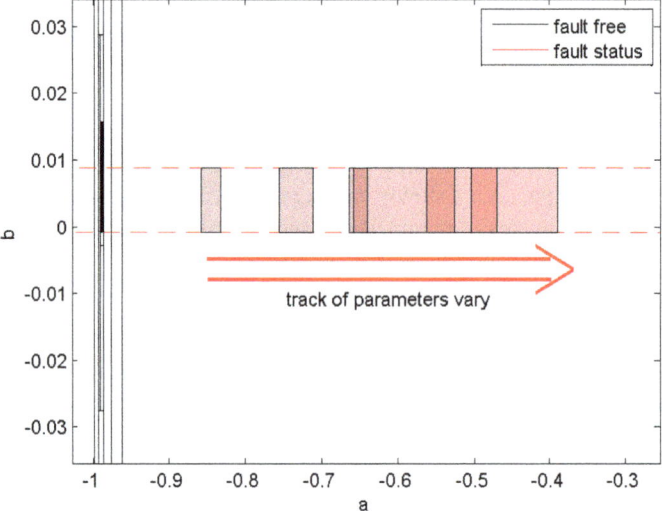

Fig. 6.28 Estimation recursive evolution process of parameters a and b versus k obtained by applying the PDE-FFD algorithm

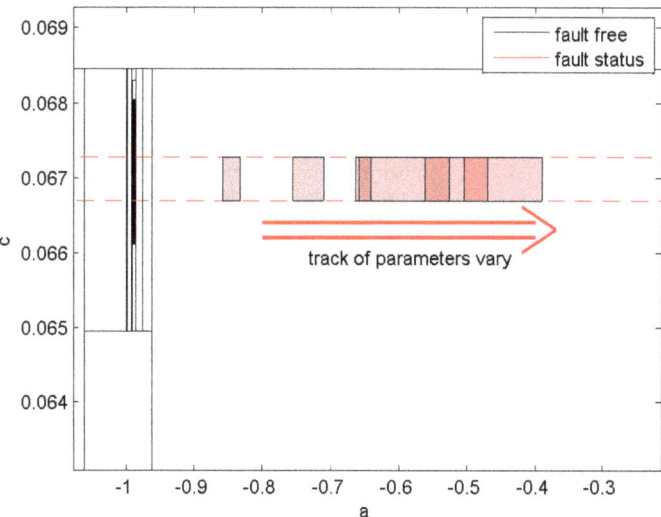

Fig. 6.29 Estimation recursive evolution process of parameters a and c versus k obtained by applying the PDE-FFD algorithm

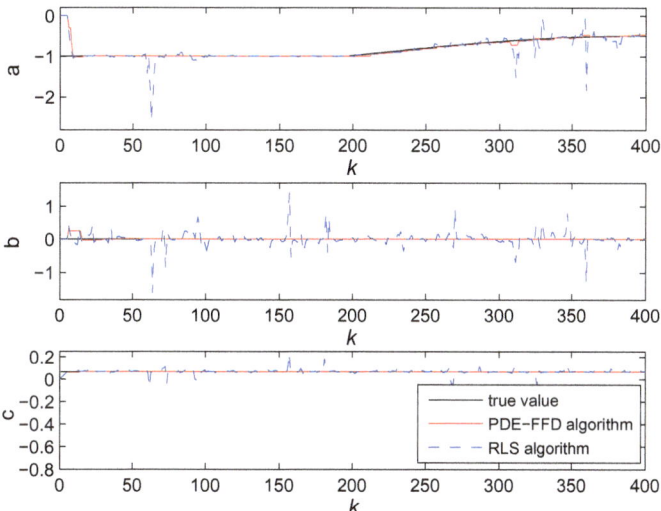

Fig. 6.30 Estimation of parameter tendency versus k obtained by applying the PDE-FFD algorithm

Fig. 6.31 Fault detection and isolation signal versus k

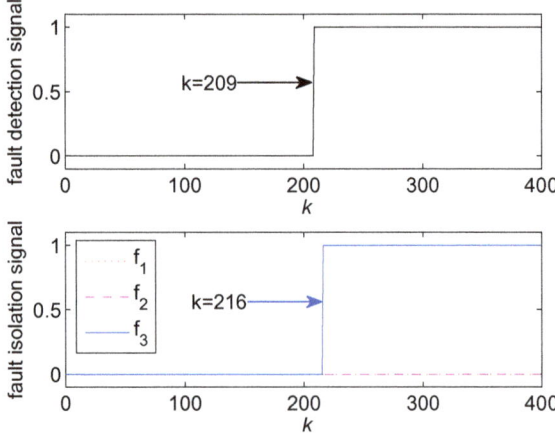

Fig. 6.32 Estimation recursive evolution process of parameters versus k

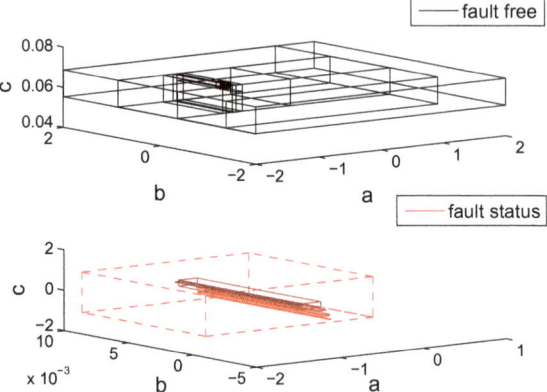

(ii) Figures,6.32, 6.33 and 6.34 show the estimation recursive evolution process of parameters in the fault-free case (black boxes) and fault status (red boxes). When the isolation process ends and it is detected that parameters a and c have failed, only parameters a and c are expanded, as shown as red dotted boxes in the three figures. During estimation of the fault state, parameter b estimates remain unchanged, and the estimated values of parameter a vary along the direction in which the a axis increases and the estimated values of parameter c vary along the direction in which the c axis decreases.

(iii) It can be seen from Fig. 6.35 that the tendency of parameter estimation is consistent with the trend or the slope of the real parameters of the system. In addition, it can be seen that, under the condition in which parameters a and c fail, there is no influence on the estimates of the fault-free parameter b. Similarly, the tracking effect of the RLS algorithm is worse than the PDE-FFD algorithm.

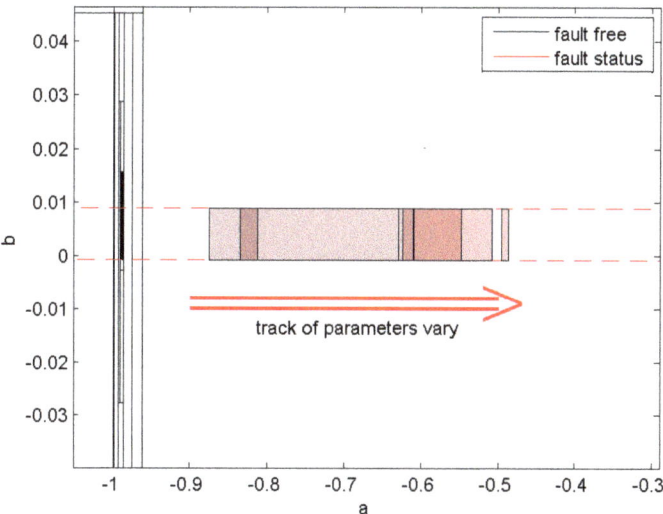

Fig. 6.33 Estimation recursive evolution process of parameters a and b versus k

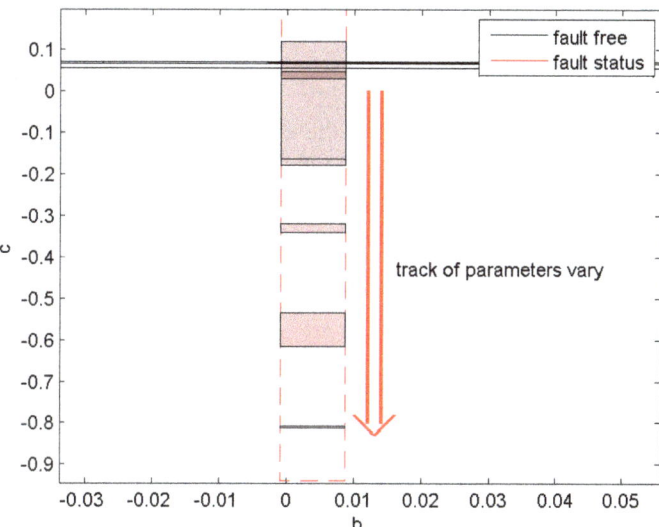

Fig. 6.34 Estimation recursive evolution process of parameters b and c versus k

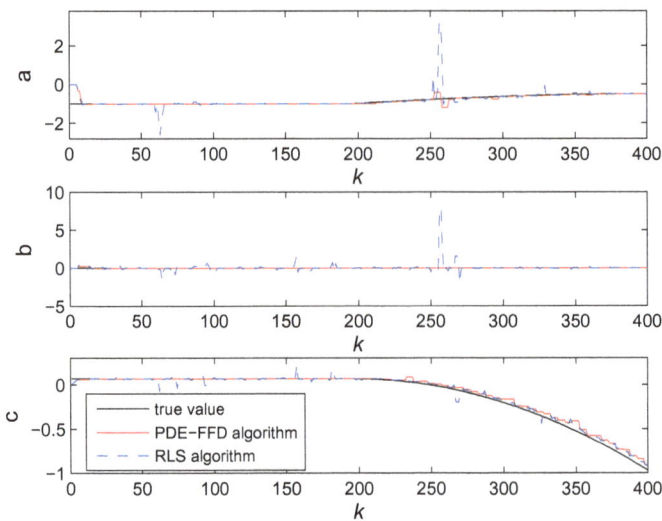

Fig. 6.35 Estimation of parameter tendency versus k

6.8 Application Study

Example 1 In order to further verify the effectiveness of the orthotopic spatial expansion filtering method in solving the identification problem of bounded time-varying parameter systems, the following uses the wind turbine pitch subsystem as a simulation example for analysis.

In the wind turbine system, the pitch subsystem is an important part of controlling the blade and pitch angle. Its structural model is [13]:

$$\begin{bmatrix} \dot{\beta} \\ \dot{\beta}_a \end{bmatrix} = \begin{bmatrix} 0 & 1 \\ -\omega_n^2 & -2\zeta\omega_n \end{bmatrix} \begin{bmatrix} \beta \\ \beta_a \end{bmatrix} + \begin{bmatrix} 0 \\ \omega_n^2 \end{bmatrix} \beta_r. \tag{6.89}$$

where, β and β_a are the pitch angle and angular velocity respectively, β_r is reference value of pitch, ζ and ω_n are the damping coefficient and the natural frequency of the system, ζ and ω_n will change with changes in hydraulic pressure, mainly due to the drop in main line pressure, the faults considered in the hydraulic system may cause dynamic changes in parameters. This dynamic change is the change of the damping coefficient between the nominal value of 0.6 rad/s and 0.9 rad/s and the change in the natural frequency of the system between 3.42 rad/s and the nominal value 11.11 rad/s [14]. Approximate the mathematical model of the pitch subsystem (6.89) to a second-order system [15]:

$$\frac{y}{u} = \frac{\omega_n^2}{s^2 + 2\zeta\omega_n s + \omega_n^2}. \tag{6.90}$$

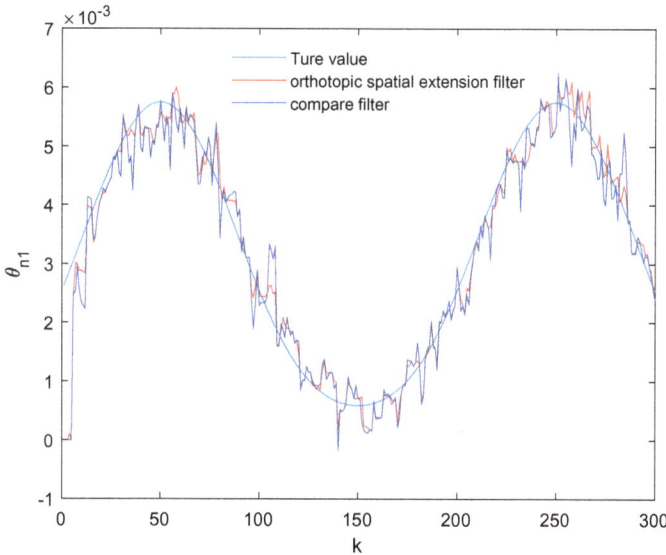

Fig. 6.36 Comparison of parameter estimates θ_{n1} when the expansion coefficients are different

where, $y = \beta$, $u = \beta_r$. In order to satisfy the dynamic changes of the parameters within the specified interval, the damping coefficient and the natural frequency change are respectively set as $\zeta = 0.75 + 0.1 \times \sin(\pi \times k \times 0.01)$ and $\omega_n = 7.265 + 3.8 \times \sin(\pi \times k \times 0.01)$. According to [13], the sampling time is $T_s = 0.01s$, and the closed-loop system is discretized as: $A(z)y(t) = B(z)u(t) + e(t)$. According to the discretization conditions, $A(z)$ and $B(z)$ are all polynomials, where $A(z) = 1 + \theta_{s1}z^{-1} + \theta_{s2}z^{-2}$, $B(z) = \theta_{n1}z^{-1} + \theta_{n2}z^{-2}$. In order to verify the effectiveness of the proposed algorithm in tracking parameter changes, in the process of parameter changes, the parameters to be identified corresponding to the $B(z)$ polynomial are analyzed and the parameter identification effect and the evolution of the orthotope are described, and the input $u(t) \in U[-25, 25]$, noise $e(t) \in U[-0.01, 0.01]$. At the same time, it is compared with the maximum value of parameter change selected as the expansion coefficient in 24, and the comparison results are shown in Figs. 6.36 and 6.37.

It can be seen from Figs. 6.36 and 6.37 that using the method in this chapter and the method proposed in [1] to solve the expansion coefficient respectively, can effectively track the time-varying parameter changes. In contrast, the filtering method proposed in this chapter based on the orthotopic spatial expansion filter has a smoother identification process and is less affected by noise disturbance, which is better than the method of selecting the maximum value of parameter changes proposed in [1] as the expansion coefficient method.

In addition, in Fig. 6.38, the interval between $k = 50 \sim 150$ is 25 to take the data of orthotopes with different expansion coefficients. The symbol x represents the true

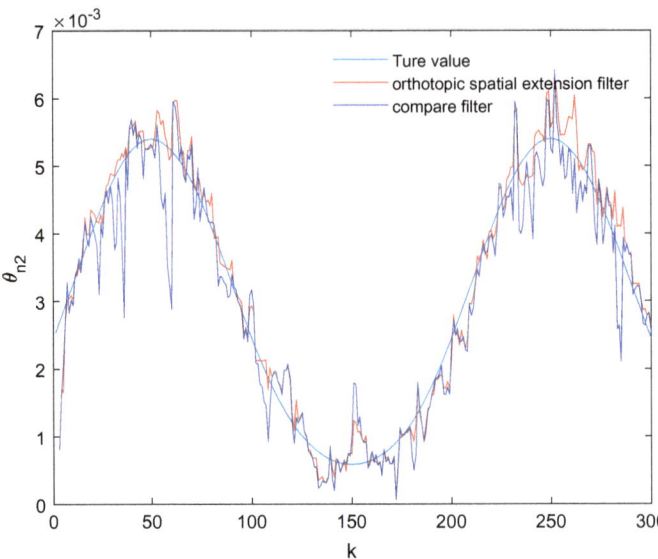

Fig. 6.37 Comparison of parameter estimates θ_{n2} when the expansion coefficients are different

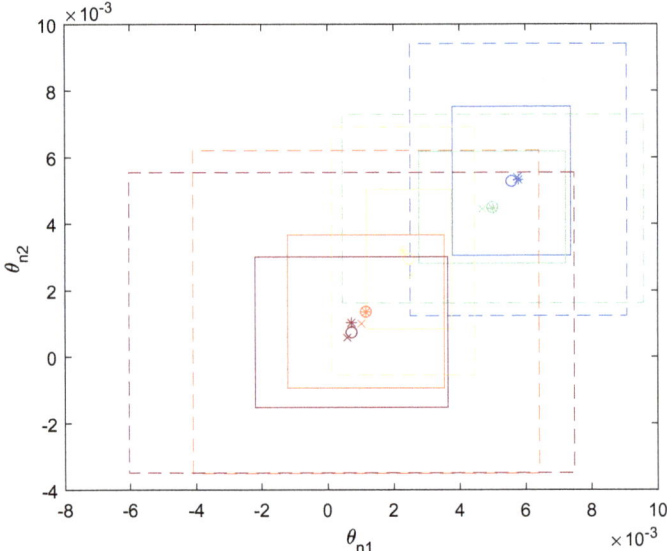

Fig. 6.38 Recursive evolution of orthotopes with different expansion coefficients

Fig. 6.39 Buck converter topology

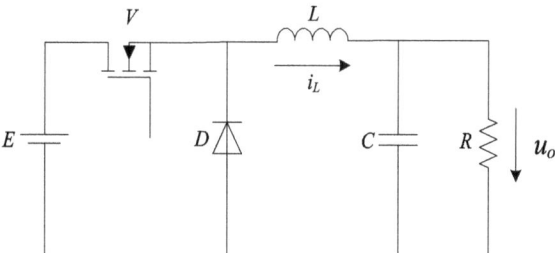

Fig. 6.40 Buck converter equivalent schematic

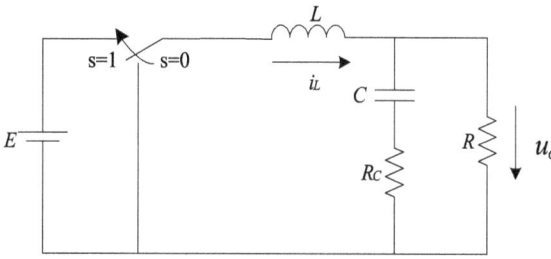

value of the parameter at the current moment. The solid line wrapping area and the symbol o respectively represent the estimated values of the linear programming expansion coefficient and the center of the orthotope at the current moment, the dashed-line wrapped area and the mark * indicate the maximum expansion coefficient of the selected parameter change at the current moment and the estimated values of the center of the orthotope. It can be seen that the orthotope based on linear programming is more compact, so the algorithm is less conservative, and the center estimate is closer to the true value of the parameter.

Example 2

The fault condition in buck circuit degradation is used as example in this experiment to evaluate the effectiveness of the orthotopic-filtering-based hierarchical fault diagnosis algorithm.

As we can see in Fig. 6.39, the power electronic circuits include some main components, such as power switches, inductors, capacitors, resistors, and power diodes. A dual-select switch s is used to represent the power diode and the controllable device. s = 1 is set if the controllable switch (MOSFET) is turned on, while, s=0 if the diode is turned on. The electrolytic capacitor is regarded as being connected in series to both the series capacitor and the equivalent series resistance. Inductance is regarded as an ideal device. There is no influence of series resistance. The obtained equivalent schematic diagram is shown in Fig. 6.40.

The hybrid system model expression for the buck converter in continuous current mode (CCM) is

$$\begin{bmatrix} \dot{i}_L \\ \dot{u}_0 \end{bmatrix} = \begin{bmatrix} 0 & -\frac{1}{L+\hat{R}_cRC} \\ \frac{R}{(R_c+R)C} & -\frac{L+\hat{R}_cRC}{(R_c+R)LC} \end{bmatrix} \begin{bmatrix} i_L \\ u_0 \end{bmatrix}$$
$$+ \begin{bmatrix} \frac{E}{L} \\ \frac{R_c\hat{R}E}{(R_c+R)L} \end{bmatrix} s. \tag{6.91}$$

Discretizing Eq. (6.91) gives

$$\begin{bmatrix} i_L(t) \\ u_0(t) \end{bmatrix} = \begin{bmatrix} 1 & -\frac{T_s}{L} \\ \frac{T_sR}{(R_c+R)C} & 1 - \frac{(L+\hat{R}_cRC)T_s}{(R_c+R)LC} \end{bmatrix} \begin{bmatrix} i_L(t-1) \\ u_0(t-1) \end{bmatrix}$$
$$+ \begin{bmatrix} \frac{ET_s}{L} \\ \frac{R_c\hat{R}ET_s}{(R_c+R)L} \end{bmatrix} s(t-1), \tag{6.92}$$

where T_s is the sampling period.

$e(t)$ and $w(t)$ are taken into account as the unknown but bounded white noises, Eq. (6.92) can be written as an identification model as

$$i_L(t) = \varphi^T(t)\theta_i + e(t), \tag{6.93}$$
$$u_0(t) = \varphi^T(t)\theta_u + w(t), \tag{6.94}$$

where $\varphi(t) = [i_L(t-1), u_0(t-1), s(t-1)]^T$, $\theta_i := [\theta_1, \theta_2, \theta_3]^T = [1, -\frac{T_s}{L}, \frac{ET_s}{L}]^T$, $\theta_u := [\theta_4, \theta_5, \theta_6]^T = [\frac{T_sR}{(R_c+R)C}, 1 - \frac{(L+R_cRC)T_s}{(R_c+R)LC}, \text{ and } \frac{R_cRET_s}{(R_c+R)L}]^T$. Then, selecting the estimates of $\hat{\theta}_3$, $\hat{theta}_4$, $\hat{\theta}_5$, and $\hat{\theta}_6$ enables us to solve for the component parameters $\hat{L} = \frac{ET_s}{\hat{\theta}_3}$, $\hat{R} = \frac{E\hat{\theta}_4}{(1-\hat{\theta}_5)E-\hat{\theta}_6}$, $\hat{R}_c = \frac{\hat{L}\hat{R}\hat{\theta}_6}{\hat{R}ET_s-\hat{L}\hat{\theta}_6}$, and $\hat{C} = \frac{T_s\hat{R}}{(\hat{R}_c+\hat{R})\hat{\theta}_4}$.

When the buck circuit works normally, the parameters are set as $L = 102 \ \mu H$, $C = 18.2 \ \mu F$, $R_c = 0.597 \ \Omega$, $R = 2.854 \ \Omega$, $T_s = 10^{-6}$ s, switching frequency $f = 50$ kHz, duty cycle $D = 0.5$, and voltage $E = 30$ V, and the noise signal consists of uniform random numbers that are randomly distributed in the interval $[-\delta, \delta]$ with $\delta = 0.01, 0.02$, and 0.03, respectively.

A simulation model is used to obtain the sampled data, and the method of orthotopic filtering is applied to realize parameter estimation of the buck circuit. The parameter estimation results are given in Table 6.2.

The following can be seen from Table 6.2.

1. At the beginning, the estimation error decreases rapidly, and the estimation error decreases slowly as the sampling time increases, indicating the good convergence of the orthotopic filtering method.

2. As the sampling time increases, the circuit parameter estimation values become closer to their true value, and the estimation error $\Delta\theta$ (%) $\leqslant 3\%$, which indicates that the method can effectively estimate the circuit parameters, demonstrating the accuracy of the algorithm.

3. The orthotopic filtering method still works with the noise boundary increasing.

The degradation of these components in the circuit can lead to parametric faults in the power electronic circuits. And we only consider the degradation of equivalent

Table 6.2 DC motor model parameters

δ	t	$\hat{L}$ (μH)	$\hat{C}$ (μF)	$\hat{R}_c$ (Ω)	$\hat{R}$ (Ω)	$\Delta\theta$ (%)
0.01	500	104.37897	15.59391	0.64785	2.74766	3.40609
	1000	104.37897	17.78277	0.64580	2.78516	2.33150
	1500	103.00975	17.99758	0.63584	2.77666	0.99677
	2000	102.72379	18.14411	0.63356	2.77883	0.70461
0.02	500	106.36903	14.42538	0.63580	2.71537	5.57196
	1000	106.36903	17.39006	0.63300	2.76763	4.28779
	1500	104.30027	17.81705	0.61878	2.75105	2.25189
	2000	103.68870	18.10500	0.61412	2.75543	1.63441
0.03	500	108.43644	13.27494	0.62265	2.69376	7.82052
	1000	108.43644	16.90368	0.61981	2.74833	6.33516
	1500	105.60103	17.60706	0.60109	2.72699	3.52293
	2000	104.65544	18.03695	0.59433	2.73140	2.56925
True values		102.00000	18.20000	0.59700	2.85000	

resistance, equivalent capacitance, and inductance in this chapter. Among them, the method for degrading the equivalent capacitance and inductance of the data acquisition reference [16] and the trend of equivalent series resistance with time can be approximated as follows [17]:

$$\frac{1}{R_c(t)} = \frac{1}{R_c(0)} \times (1 - K \times t \times e^{-\frac{4700}{T+273}}), \tag{6.95}$$

where T is the internal temperature of the capacitor, $R_c(t)$ is the R_c value at time t, $R_c(0)$ is the R_c value of the capacitor at the initial moment, and K is a constant. Moreover, according to the literature, R_c is $2.8 \cdot R_c(0)$ at the end of its life. In this chapter, we take $T = 25$ and $K = 972.7275$. In addition, the noise $e(t) \in U[-0.01, 0.01]$ is only taken in the case of a fault.

In actual situations, all the components in buck circuit often appear to be degraded gradually at the same time. With the increase in the use time of power electronic circuits, the trends of component degradation are as follows: R_c increases and C and L decrease over time. The component parameter values of the circuit at 25 time points are selected as 25 fault types to form a fault library. The specific fault types are listed in Table 6.3.

Further more, in order to highlight the superiority of the fault diagnosis methods proposed in this chapter, a model matching fault diagnosis method is used for comparison. For each fault type FT(1) $\sim$ FT(25) and new fault type, a corresponding fault identification filter $f_m(1) \sim f_m(25)$ and the corresponding fault type filters are designed.

Take circuit failure 15 as an example. The fault type FT(15) is added at time $t = 2000$ when the circuit is working normally. In this simulation, let $\alpha = 1.0, \beta = 3.5$,

Table 6.3 DC motor model parameters

No.	True values				Estimated values			
	L (μH)	C (μF)	R_c (Ω)	R (Ω)	$\hat{L}$ (μH)	$\hat{C}$ (μF)	$\hat{R}_c$ (Ω)	$\hat{R}$ (Ω)
FT(0)	102	18.2	0.597	2.85	102.724	18.138	0.634	2.779
FT(1)	101.958	18.153	0.612	2.85	100.861	17.741	0.698	2.854
FT(2)	101.466	17.886	0.628	2.85	100.396	17.482	0.714	2.853
FT(3)	100.924	17.599	0.645	2.85	99.885	17.205	0.732	2.852
FT(4)	100.333	17.295	0.663	2.85	99.327	16.911	0.751	2.850
FT(5)	99.692	16.973	0.681	2.85	98.721	16.600	0.770	2.849
FT(6)	99.003	16.632	0.701	2.85	98.071	16.270	0.791	2.847
FT(7)	98.267	16.274	0.722	2.85	97.375	15.923	0.814	2.845
FT(8)	97.483	15.898	0.744	2.85	96.634	15.559	0.837	2.844
FT(9)	96.652	15.504	0.768	2.85	95.847	15.177	0.863	2.842
FT(10)	95.775	15.093	0.793	2.85	95.017	14.778	0.889	2.840
FT(11)	94.852	14.665	0.820	2.85	94.142	14.362	0.918	2.838
FT(12)	93.883	14.220	0.849	2.85	93.223	13.930	0.950	2.836
FT(13)	92.870	13.758	0.880	2.85	92.261	13.481	0.983	2.834
FT(14)	91.813	13.280	0.914	2.85	91.257	13.015	1.020	2.832
FT(15)	90.712	12.785	0.950	2.85	90.209	12.533	1.059	2.830
FT(16)	89.568	12.273	0.989	2.85	89.120	12.033	1.101	2.828
FT(17)	88.381	11.745	1.031	2.85	87.989	11.518	1.147	2.826
FT(18)	87.153	11.202	1.077	2.85	86.819	10.987	1.198	2.824
FT(19)	85.882	10.642	1.127	2.85	85.607	10.439	1.253	2.821
FT(20)	84.571	10.067	1.182	2.85	84.356	9.876	1.314	2.819
FT(21)	83.219	9.476	1.243	2.85	83.064	9.296	1.381	2.817
FT(22)	81.827	8.870	1.311	2.85	81.734	8.702	1.457	2.815
FT(23)	80.396	8.248	1.386	2.85	80.365	8.092	1.541	2.813
FT(24)	78.925	7.611	1.471	2.85	78.956	7.468	1.637	2.812
FT(25)	77.417	6.960	1.566	2.85	77.511	6.829	1.745	2.810

$\gamma = 0.077, \sigma = 0.9, \varepsilon = 0.03, l = 10, Q = 10$, and $T_N = 500$. Applying the method proposed in this chapter for fault diagnosis gives the results shown in Figs. 6.41~6.43.

As shown in Fig. 6.41, the fault library is divided into six layers through hierarchical clustering analysis. The clustering result is used as the prior knowledge of the next fault diagnosis analysis, and a discriminant analysis is performed layer by layer.

From Figs. 6.42 and 6.43, the following can be seen:

1. The fault detection signal jumps at time $t = 2001$, which means that a circuit failure is detected at time $t = 2001$. The proposed method can detect and display the fault in time. After the fault is detected, a discriminant analysis is performed layer by layer according to the hierarchical clustering analysis result shown in Fig. 6.41 until

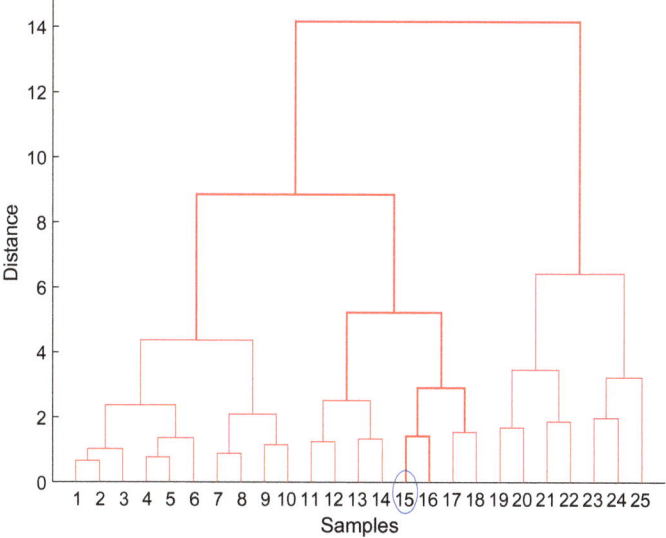

Fig. 6.41   Hierarchical cluster analysis result of fault library

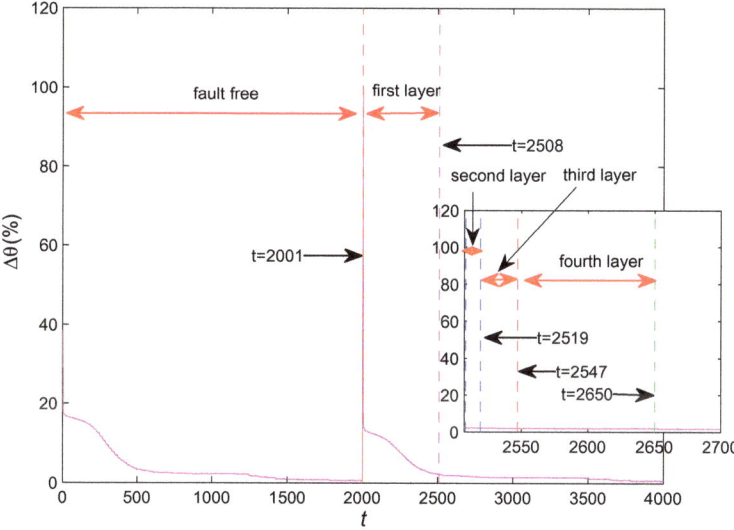

Fig. 6.42   Estimation errors $\Delta\theta(\%)$ versus t

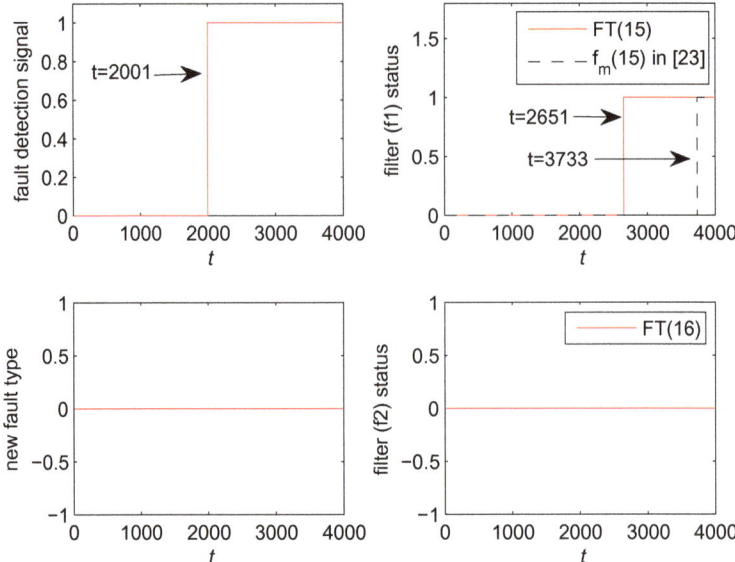

Fig. 6.43 Fault diagnosis result

the number of samples in the layer is $\leqslant 2$. Then we can determined the first, second, third, and fourth layers at times $t = 2508, 2519, 2547$, and 2650, respectively.

2. When the number of samples in the layer is $\leqslant 2$, the types of faults included are FT(15) and FT(16). By applying the fault-matching method to realize fault identification, $f1$ jumps at time $t = 2651$, indicating that the fault type is identified as FT(15) at time $t = 2651$; at this time, the filter state of the new fault type is not hopped, indicating that the fault is included in the fault library; i.e., no new fault types are generated.

3. The fault detection rapidity and the fault identification accuracy are usually used as the performance evaluation indexes of fault diagnosis and identification processes, respectively. The algorithm proposed in this chapter can quickly implement fault detection. After detecting that a fault occurs in the system, the algorithm can realize fault identification within a certain time due to the influence of system noise and estimation errors, which shows the effectiveness and feasibility of the algorithm.

4. It can be seen from Fig. 6.43 that, due to there are many types of faults and the values of each fault type are very close, the fault diagnosis time is longer, i.e., the fault type is identified as FT(15) at time $t = 3733$ obtained by the model matching method in [18]. Therefore, compared with the model matching method, the fault diagnosis method based on the orthotopic-filtering-based hierarchical fault diagnosis algorithm proposed in this chapter has the advantages of lower time complexity and higher fault identification efficiency.

Add a new fault type to the circuit at time $t = 2000$, i.e., $L = 75.870 \ \mu H$, $C = 6.294 \ \mu F$, $R_c = 1.675 \ \Omega$, and $R = 2.85 \ \Omega$. In this simulation, let $\alpha = 1.0$, $\beta = 3.5$,

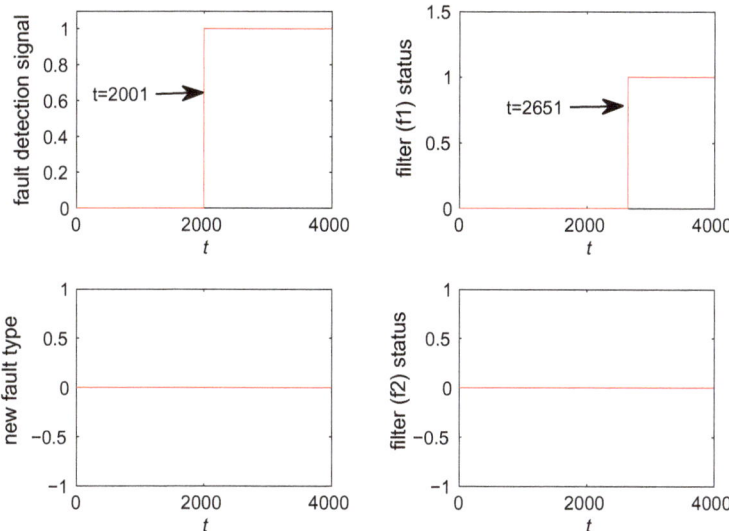

Fig. 6.44   Fault diagnosis result

$\gamma = 0.077, \sigma = 0.9, \varepsilon = 0.01, l = 10, Q = 10,$ and $T_N = 500.$ Applying the method proposed in this chapter for fault diagnosis gives the results shown in Fig. 6.19.

The fault detection signal jumps at time $t = 2001$ in Fig. 6.44, indicating that a fault is detected at time $t = 2001.$ However, the states of the identified filters $f1$ and $f2$ are jump free, and the new fault type filter hops, indicating that the fault type is not in the fault library and so the circuit generates a new fault type.

The estimated values of the fault parameters are obtained by using the above fault diagnosis process, i.e., $\hat{L} = 76.026\ \mu\text{H}, \hat{C} = 6.177\ \mu\text{F}, \hat{R}_c = 1.870\ \Omega,$ and $\hat{R} = 2.809\ \Omega.$ The fault type is numbered as FT(26) and added to the fault library. Then, fault FT(26) is added again at time $t = 2000.$ Using the hierarchical fault diagnosis algorithm based on the orthotopic filtering described in this chapter achieves fault diagnosis. The fault diagnosis results are shown in Figs. 6.45~6.47.

It can be seen from Fig. 6.45 that hierarchical clustering analysis of the fault library is performed again after adding the fault library sample, and the clustering result is used as the prior knowledge of the next fault diagnosis analysis.

From Figs. 6.46 and 6.47, the following can be seen:

1. A circuit failure is detected at time $t = 2001$ according to the fault detection signal jumps at time $t = 2001.$ Then, a discriminant analysis is performed layer by layer according to the hierarchical clustering analysis result shown in Fig. 6.45 until the number of samples in the layer is $\leqslant 2.$ Then the first, second, and third layers at times $t = 2457, 2562,$ and $2563,$ respectively, are determined.

2. When the number of samples in the layer is $\leqslant 2,$ the types of faults included are FT(25) and FT(26). After fault identification using the fault-matching method , $f2$ jumps at time $t = 2564,$ indicating that the fault type is identified as FT(26) at time

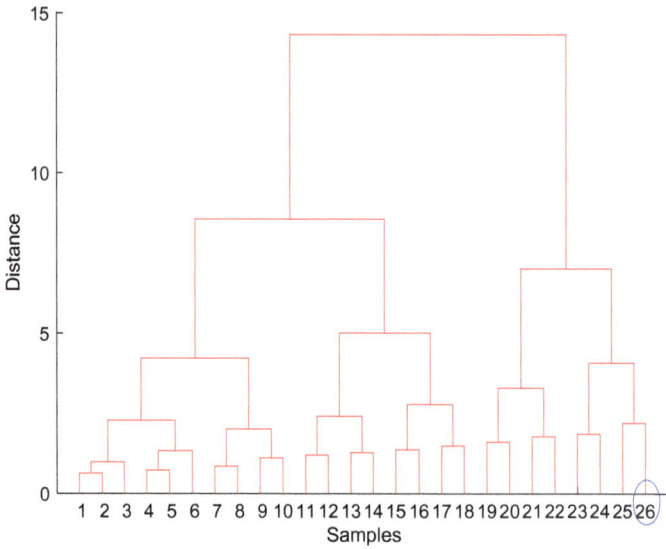

Fig. 6.45 Hierarchical cluster analysis result of fault library

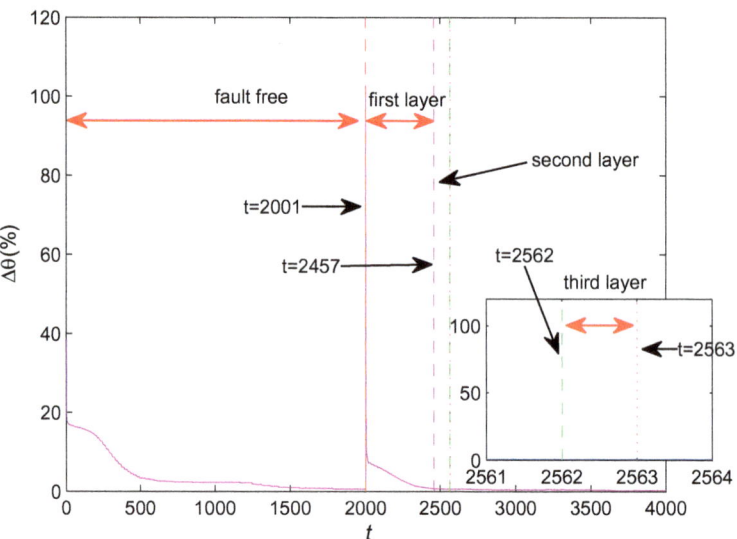

Fig. 6.46 Estimation errors $\Delta\theta(\%)$ versus t

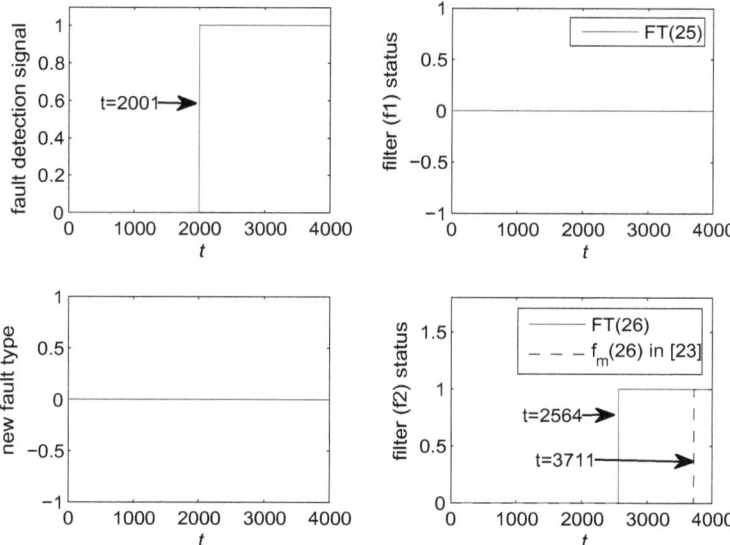

Fig. 6.47 Fault diagnosis result

$t = 2564$, and it is quicker then adopting the model matching algorithm, whose fault identification time is $t = 3711$. Therefore, we can draw that the method proposed in this chapter has better fault diagnosis effect. Additionally, the filter state of the new fault type is not hopped, indicating that the fault is included in the fault library.

6.9 Concluding Remarks

A parameter global expansion filtering fault diagnosis algorithm and a parameter direction expansion filtering fault diagnosis algorithm based on spatial dimension reduction for IN-CAR systems have been presented. First, the running status of the current system (i.e., whether the parameter feasible set is empty) is detected. When the feasible set of the parameter is a null set, it is determined that the system is in a faulty state; otherwise, the system is in a normal state. Second, after detecting the system condition, the parameter global expansion filtering fault diagnosis algorithm and the spatial-dimension-reduction-based parameter directional expansion filtering fault diagnosis algorithm are, respectively, proposed to capture the trend of the parameter change under the current slowly varying faults. Finally, different types of slowly varying faults are determined by the slope and trend of the parameter estimation curve.

The given fault diagnosis algorithms are effective and can identify different fault types, even faults that are slowly varying. Consequently, the studied PGE-FFD and PDE-FFD algorithms can be used in the area of nonlinear parameter estimation, mechanical fault detection, high-precision testing, and other high-sensitivity industrial process analysis. The orthotopic-filtering-based fault diagnosis algorithms for nonlinear systems with slowly varying faults can combine other multi-innovation algorithms [19, 20] and the recursive algorithms [21–23] to study the parameter identification of other linear and nonlinear stochastic systems woth colored noises and can be applied to other literatures, such as engineering application systems [24–28].

We have presented an orthotopic-filtering-based hierarchical fault diagnosis algorithm. The proposed method is used to envelop the feasible parameter set, and the orthotopic recursive solution process is described by solving the problem of finite linear programming. Whether the fault occurs is judged by determining whether the feasible parameter set is empty. When the system works normally, the feasible parameter set must not be empty. When the feasible parameter set is empty, the system must have failed. The hierarchical clustering method is applied to cluster analysis of the fault database, and the clustering result is used as the prior knowledge of the next fault diagnosis analysis. According to the clustering result, a discriminant analysis is performed layer by layer until the fault sample contained in the layer is no more than two. The parameter sequence obtained by the analysis and the model-matching method are used for fault identification. The parameter degradation problem of a buck circuit is taken as an example to verify the effectiveness and feasibility of the algorithm.

The fault diagnosis method proposed in this chapter is also applicable to other industrial process fault diagnosis problems with unknown but bounded noise disturbance, such as the robot fault detection [29, 30] and other research fields based on filtering [31, 32].

References

1. M. Casini, A. Garulli, and A. Vicino. A constraint selection technique for set membership estimation of time-varying parameter, in *Proceedings of 53rd IEEE Conference on Decision and Control*, vol. 32, pp. 1029–1034 (2014)
2. Z.Y. Wang, Z.C. Ji, Data-filtering-based iterative identification methods for nonlinear FIR-MA systems. J. Vib. Control **20**(14), 2193–2201 (2014)
3. F. Ding, Two-stage least squares based iterative estimation algorithm for CARARMA system modeling. Appl. Math. Model. **37**(7), 4798–4808 (2013)
4. F. Ding, Decomposition based fast least squares algorithm for output error systems. Signal Process. **93**(5), 1235–1242 (2013)
5. F. Ding, Y. Liu, B. Bao, Gradient-based and least-squares-based iterative estimation algorithms for multi-input multi-output systems. P. I. Mech. Eng. C. J. Mech. **226**(1), 43–55 (2012)
6. Z.Y. Wang, Y.X. Shen, D.H. Wu, Hierarchical least squares algorithms for nonlinear feedback system modeling. J. Frankl. Inst. **353**(10), 2258–2269 (2016)
7. X.K. Wan, Y. Li, C. Xia, A T-wave alternans assessment method based on least squares curve fitting technique. Meas. **86**, 93–100 (2016)

8. J. Pan, X. Jiang, X.K. Wan, W.F. Ding, A filtering based multi-innovation extended stochastic gradient algorithm for multivariable control systems. Int. J. Control Autom. Syst. **15**(3), 1189–1197 (2017)
9. J. Pan, W. Li, H.P. Zhang, Control algorithms of magnetic suspension systems based on the improved double exponential reaching law of sliding mode control. Int. J. Control Autom. Syst. **16**(6), 2878–2887 (2018)
10. R.R. Zheng, Mao Z.Y., and X.X. Luo. Artificial immune algorithm based on Euclidean distance and king-crossover. Control and Decis. **20**(2), 161–164 (2005)
11. J. Blesa, V. Puig, J. Saludes, Robust fault detection using polytope-based set-membership consistency test. IET Control Theory & Appl. **6**(12), 1767–1777 (2012)
12. H.Y. Hu, R. Ding, Least squares based iterative identification algorithms for input nonlinear controlled autoregressive systems based on the auxiliary model. Nonlinear Dyn. **76**(1), 777–784 (2013)
13. P. Casau, P. Rosa, S.M. Tabaeipour, A set-valued approach to FDI and FTC of wind turbines. IEEE Trans. Control Syst. Technol. **13**(1), 245–263 (2014)
14. C. Sloth, T. Esbensen, J. Stoustrup, Robust and fault-tolerant linear parameter-varying control of wind turbines. Mechatronics **21**(4), 645–659 (2011)
15. S.M. Tabatabaeipour, Active fault detection and isolation of discrete-time linear time-varying system: a set-membership approach. Int. J. Syst. Sci. **46**(11), 1917–1933 (2015)
16. Y.Y. Jiang, Y.R. Wang, H. Luo, An innovative metric for power electronic circuit failure evaluation and a novel prediction method based on LSSVM. T. China El. Soc. **27**(12), 43–50 (2012)
17. S.Y. Wang, D.J. Chen, Y.Z. Ye, Research on fault prediction method of power electronic circuits based on particle swarm optimization RBF neural network. Appl. Mech. Mater. **687–691**, 3354–3360 (2014)
18. M. Hashimoto, F. Itaba, K. Takahashi, Model-based fault detection and isolation for a powered wheelchair, in *Proceedings of the 4th International Symposium on Communications*, vol. 32, pp. 1–6 (2010)
19. F. Ding, X. Zhang, L. Xu, The innovation algorithms for multivariable state-space models. Int. J. Adapt. Control Signal Process. **33**(11), 1601–1618 (2019)
20. F. Ding, Hierarchical multi-innovation stochastic gradient algorithm for Hammerstein nonlinear system modeling. Appl. Math. Model. **37**(4), 1694–1704 (2013)
21. J. Ding, F. Ding, X.P. Liu, G.j. Liu.,Hierarchical least squares identification for linear SISO systems with dual-rate sampled-data. IEEE T. Automat. Contr. **56**(11), 2677–2683 (2011)
22. F. Ding, Coupled-least-squares identification for multivariable systems. IET Control Theory Appl. **7**(1), 68–79 (2013)
23. Z.Y. Wang, Y.X. Shen, Z.C. Ji, R. Ding, Filtering based recursive least squares algorithm for Hammerstein FIR-MA systems. Nonlinear Dyn. **73**(1–2), 1045–1054 (2013)
24. Y. Gu, J.C. Liu, X.L. Li, Y.X. Chou, Y. Ji, State space model identification of multirate processes with time-delay using the expectation maximization. J. Frankl. Inst. **356**(3), 1623–1639 (2019)
25. Y. Gu, Y.X. Chou, J.C. Liu, Y. Ji, Moving horizon estimation for multirate systems with time-varying time-delays. J. Frankl. Inst. **356**(4), 2325–2345 (2019)
26. X.L. Zhao, Z.Y. Lin, B. Fu, L. He, N. Fang, Research on automatic generation control with wind power participation based on predictive optimal 2-degree-of-freedom PID strategy for multi-area interconnected power system. Energies **11**(12), 3325 (2018)
27. N. Zhao, Y.C. Liang, Y.Y. Pei, Dynamic contract incentive mechanism for cooperative wireless networks. IEEE T. Veh Technol. **67**(11), 10970–10982 (2018)
28. N. Zhao, Y.C. Liang, Deep Reinforcement Learning for User Association and Resource Allocation in Heterogeneous Cellular Networks. IEEE T. Wirel. Commun. **18**(11), 5141–5152 (2019)
29. S.J. Yoo, Actuator fault detection and adaptive accommodation control of flexible-joint robots. IET Control Theory Appl. **6**(10), 1497–1507 (2012)
30. X.J. Li, G.H. Yang, Robust fault detection and isolation for a class of uncertain single output non-linear systems. IET Control Theory Appl. **8**(7), 462–470 (2014)

31. Y.J. Wang, F. Ding, Iterative estimation for a non-linear IIR filter with moving average noise by means of the data filtering technique. IMA J. Math. Control I. **34**(3), 745–764 (2017)
32. L. Li, L. Yang, Z.L. Yang, Mean-square error constrained approach to robust stochastic iterative learning control. IET Control Theory Appl. **12**(1), 38–44 (2017)

Chapter 7
Fault Diagnosis Method Based on Composite Set-Membership Filter

7.1 Preliminaries and Problem Formulation

A linear discrete-time model is considered as follow:

$$y(k) = \phi^{\mathrm{T}}(k)\theta + e(k), k = 1, 2, \ldots, N, \tag{7.1}$$

where $\phi(k) \in \mathbb{R}^n$, $\theta \in \mathbb{R}^n$, $u(k)$ and $y(k)$ are the observable regression vector, the unknown parameter vector, the input and output data of the system, respectively, and $e(k)$ is the unknown but bounded noise sequence bounded by γ.

$$\phi(k) = [-y(k-1), \ldots, -y(k-n_a), u(k), \ldots, u(k-n_b)]^{\mathrm{T}},$$
$$\theta = [a_1, \ldots, a_{n_a}, b_0, \ldots, b_{n_b}]^{\mathrm{T}},$$
$$|e(k)| \leqslant \gamma, \gamma \geqslant 0, \tag{7.2}$$

where $n = n_a + n_b + 1$ is the total number of parameters. Based on the measuring data and known noise bounds, we can use the set membership estimation to determine the feasible parameter set. Therefore, the measurement set $S(k)$ can be defined as

$$S(k) = \{\theta : |y(k) - \phi^{\mathrm{T}}(k)\theta| \leqslant \gamma\}. \tag{7.3}$$

From Eq. (7.3), we can get that all the possible parameters of the system at time k are contained between two parallel hyperplanes. The feasible parameters of the system are contained in a convex polytopic set with the increase in sample time. An ellipsoid is used to outer bound the convex polytopic set for simplicity as it has lesser computational complexity than a zonotope, a polytope, and an orthotope. We define the ellipsoid $E(k)$ as

$$E(k) = \{\theta : (\theta - \theta^c(k))^{\mathrm{T}} P^{-1}(k)(\theta - \theta^c(k)) \leqslant 1\}. \tag{7.4}$$

Z. Wang et al., *Advances in Fault Detection and Diagnosis Using Filtering Analysis*,
https://doi.org/10.1007/978-981-16-5959-1_7

Here, $\theta^c(k)$ is the center of $E(k)$, and $P(k)$ is the symmetric positive definite matrix describing the shape of $E(k)$ at time k.

The minimum volume ellipsoids can be obtain by bounding the convex polytopes tightly through the OVE algorithm with direct geometric meaning and easy implementation. The OVE algorithm is summarized in Algorithm 7.1 and the updating principle of the ellipsoid in [1] follows the following algorithm.

Algorithm 7.1 OVE algorithm

1: $\underline{\alpha}(k) \leftarrow \max(\frac{y(k)-\phi^T(k)\theta(k-1)-\gamma}{(\phi^T(k)P(k-1)\phi(k))^{1/2}}, -1)$

2: $\overline{\alpha}(k) \leftarrow \min(\frac{y(k)-\phi^T(k)\theta(k-1)+\gamma}{(\phi^T(k)P(k-1)\phi(k))^{1/2}}, 1)$

3: **if** $\underline{\alpha}(k) \geqslant 1$ or $\overline{\alpha}(k) \leqslant -1$ **then**

4:　　Fault indication signal $f(k) \leftarrow 1$

5:　　$\theta^c(k) \leftarrow \theta^c(k-1), P(k) \leftarrow P(k-1)$

6: **else**

7:　　Fault indication signal $f(k) \leftarrow 0$

8:　　$\varepsilon(k) \leftarrow \underline{\alpha}(k)\overline{\alpha}(k)$

9:　　**if** $\varepsilon(k) \leqslant -\frac{1}{n}$ **then**

10:　　　$\theta^c(k) \leftarrow \theta^c(k-1), P(k) \leftarrow P(k-1)$

11:　　**else**

12:　　　$\mu(k) \leftarrow \frac{\underline{\alpha}(k)+\overline{\alpha}(k)}{2}$

13:　　　**if** $|\mu(k)| > \rho$ **then**

14:　　　　$b(k) \leftarrow 2n\mu(k) + \frac{1+\varepsilon(k)}{\mu(k)}$

15:　　　　$\tau(k) \leftarrow \frac{b(k)-\text{sign}(\mu(k))\sqrt{b^2(k)-4(n+1)(1+n\varepsilon(k))}}{2(n+1)}$

16:　　　　$\sigma(k) \leftarrow \tau(k)(\tau(k) - \frac{1+\varepsilon(k)}{\mu(k)}) + 1$

17:　　　　$\zeta(k) \leftarrow \frac{\sigma(k)}{1-\frac{\tau(k)}{\mu(k)}}$

18:　　　**else**

19:　　　　$\alpha(k) \leftarrow \max(|\underline{\alpha}(k)|, |\overline{\alpha}(k)|)$

20:　　　　$\tau(k) \leftarrow 0$

21:　　　　$\sigma(k) \leftarrow n\alpha^2$

22:　　　　$\zeta(k) \leftarrow \frac{n(1-\alpha^2(k))}{n-1}$

23:　　　**end if**

24:　　　$\theta(k) \leftarrow \theta(k-1) + \frac{\tau(k)P(k-1)\phi(k)}{(\phi^T(k)P(k-1)\phi(k))^{1/2}}$

25:　　　$P(k) \leftarrow \zeta(k)P(k-1) + (\sigma(k) - \zeta(k))\frac{P(k-1)\phi(k)\phi^T(k)P(k-1)}{\phi^T(k)P(k-1)\phi(k)}$

26:　　**end if**

27: **end if**

28: **return** $\theta(k), P(k)$ and $f(k)$

$$E(k) = \arg[\min_{E}\{\text{vol}(E) : E \supset E(k-1) \cap S(k)\}] \tag{7.5}$$

The support function of the ellipsoid $E(k)$ in the direction q is defined as

$$\begin{cases} \eta(q|E(k))^+ = \langle q, \theta^c(k)\rangle + \langle q, P(k)q\rangle^{1/2}, \\ \eta(q|E(k))^- = \langle q, \theta^c(k)\rangle - \langle q, P(k)q\rangle^{1/2}, \end{cases} \tag{7.6}$$

where $q \in \mathbb{R}^n$, $\langle \cdot \rangle$ represents the inner product function and $\eta(q|E(k))^+$ and $\eta(q|E(k))^-$ are the upper and lower bounds of $E(k)$, respectively.

$O(k)$ is defined as the support orthotope of the ellipsoid $E(k)$, and its coordinates of the vertices correspond to the parameter bounds:

$$O(k) = \left\{ \theta : \begin{bmatrix} \theta^-_{o_1}(k) \\ \vdots \\ \theta^-_{o_n}(k) \end{bmatrix} \leqslant \theta \leqslant \begin{bmatrix} \theta^+_{o_1}(k) \\ \vdots \\ \theta^+_{o_n}(k) \end{bmatrix} \right\} \tag{7.7}$$

with

$$\begin{cases} \theta^+_{o_u}(k) = \eta(l_u|E(k))^+, \ u \in \{1, 2, \ldots, n\}, \\ \theta^-_{o_u}(k) = \eta(l_u|E(k))^-, \ u \in \{1, 2, \ldots, n\}, \end{cases} \tag{7.8}$$

where $l_u = [\underbrace{0, \ldots, 0, 1, 0, \ldots, 0}_{u}]^{\mathrm{T}}, u \in \{1, 2, \ldots, n\}$.

It can be seen that the support orthotope $O(k)$ is parallel to the parameter axis, with the increase in sample time, its volume is decreasing, while there is no guarantee for the convergence of its parameter bounds. Therefore, to ensure the monotonic convergence of parameter bounds, the intersection of the support orthotopes is defined as $X(k)$, which is also an orthotope:

$$X(k) = \bigcap_{t=1}^{k} O(t) = X(k-1) \cap O(k). \tag{7.9}$$

$X(k) \subseteq X(k-1)$ can be obtained in this way, and $\theta^+_{x_u}(k)$, $\theta^-_{x_u}(k)$ and the center $\theta^c_{x_u}(k)$ of the interval $[\theta^-_{x_u}(k), \theta^+_{x_u}(k)]$ can be calculated as follows:

$$\begin{aligned} \theta^+_{x_u}(k) &= \min(\theta^+_{x_u}(k-1), \theta^+_{o_u}(k)), \\ \theta^-_{x_u}(k) &= \max(\theta^-_{x_u}(k-1), \theta^-_{o_u}(k)), \\ \theta^c_{x_u}(k) &= \frac{\theta^+_{x_u}(k) + \theta^-_{x_u}(k)}{2}. \end{aligned} \tag{7.10}$$

Consider the following linear discrete state space system:

$$\begin{cases} x(k+1) = Ax(k) + Bu(k) + w(k), \\ y(k) = Cx(k) + v(k). \end{cases} \tag{7.11}$$

where $x(k) \in \mathbb{R}^{n_x}$ and $y(k) \in \mathbb{R}^{n_y}$ are respectively kth system status and measured output value, A, B, C are known matrices, $w(k) \in \mathbb{R}^{n_x}$ means state disturbance, $v(k) \in \mathbb{R}^{n_y}$ means measurement noise. Assuming that the system disturbance, measurement noise and initial state are all bounded, $w(k) \in W, v(k) \in V, x(0) \in X(0)$, where $W, V, X(0)$ are all orthotopes.

Since it is very difficult to calculate the accurate state set $X(k)$, in practice, the accurate feasible set is usually wrapped in a regular space structure, thereby reducing the computational complexity of the state estimation algorithm, but the conservativeness has also increased accordingly. This chapter proposes a state estimation algorithm based on orthotopic double filtering, which solves $2n_x$ linear programming conditions to obtain the upper and lower bounds of the state at each time, which not only reduces the computational complexity, but also effectively improves the conservativeness.

A orthotope $\mathcal{O}$ describing the feasible set of approximate parameters:

$$\mathcal{O}(\overline{x}, d) = \{x : x = \overline{x} + \text{diag}(d)m, \|m\| \leq 1\}, \tag{7.12}$$

among them, $\overline{x}, d, m \in \mathbb{R}^n, d_i \geq 0, i = 1, \ldots, n_x$; $\text{diag}(d)$ is a diagonal matrix whose diagonal value is equal to d, sets $\mathcal{F}_i = \{x \in \mathcal{O} : x_i = \overline{x}_i + d_i\}$ and $\mathcal{F}_{i+n_x} = \{x \in \mathcal{O} : x_i = \overline{x}_i - d_i\}$ are pairs of structural planes of orthotope $\mathcal{O}$.

It can be seen that the orthotope is determined by the constraint conditions $X(0)$ and $X\{y(k)\}$, by solving $2n_x$ linear programming equations, the smallest orthotope $\mathcal{O}^*(X)$ wrapping the feasible parameter set is obtained:

$$\beta_i(k) = \max e_i^T x,$$
$$\text{s.t.}$$
$$x \in X(k). \tag{7.13}$$

$$\beta_{i+n_x}(k) = \min e_i^T x,$$
$$\text{s.t.}$$
$$x \in X(k). \tag{7.14}$$

$$X(k) = \bigcap_{i=1}^{2n_x} X_i(k). \tag{7.15}$$

where, $X(k)$ is the intersection of the feasible solution sets of $2n_x$ constraints, and e_i represents the i column of the n_x-dimensional identity matrix, $i = 1, 2, \ldots, n_x$. By solving the linear programming equation, the upper and lower bounds of each parameter at step k can be obtained, and the most compact orthotope $\mathcal{O}^*(X) = \mathcal{O}(\overline{x}^*, d^*)$, where

$$\overline{x}_i^* = \frac{\beta_i(k) + \beta_{i+n_x}(k)}{2}, \tag{7.16}$$

$$d_i^* = \frac{\beta_i(k) - \beta_{i+n_x}(k)}{2}. \tag{7.17}$$

Suppose there is a zonotope $Z(p, H) = p \oplus H\mathbf{B}^n$, H is a diagonal matrix, and the diagonal elements are h_i, $i = 1, 2, \ldots, n$. If there is $p = \overline{x}$, $h_i = d_i$, then $Z(p, H) = \mathcal{O}(\overline{x}, d)$.

Suppose there is a zonotope $Z(p, H) = p \oplus H\mathbf{B}^n$ and a orthotope $\mathcal{O}(p, d)$, If there is $d_i = \sum_{j=1}^{n_H} \|H_{ij}\|$, where n_H is the number of columns of matrix H, then $\mathcal{O}(p, d)$ is the most compact orthotope wrapping $Z(p, H) = p \oplus H\mathbf{B}^n$.

7.2 Directional Expansion Based Fault Diagnosis Algorithm Using Orthotopic and Ellipsoidal Filtering

The process of parameter estimation can be summarized as follows:

1. Initialize $E(0)$ and $X(0) = O(0)$.
2. Update $E(k)$ based on $E(k-1)$ and $S(k)$.
3. Calculate the support orthotope $O(k)$ and the intersection of the support orthotopes $X(k)$.

When faults occur, the feasible set of the parameter vector is empty. So $E(k-1)$, $X(k-1)$, and $S(k)$ are reset to $E^r(k-1)$, $X^r(k-1)$, and $S^r(k)$, respectively, to contain the varied parameter vector $\theta(k^{f_m})$:

$$\theta(k^{f_m}) = \theta^o + \Delta\theta(k^{f_m}) \tag{7.18}$$

with

$$|\Delta\theta_u(k^{f_m})| \leqslant \delta_u^{\max}, u \in \{1, \ldots, n\}, \tag{7.19}$$

where $\theta^o + \Delta\theta(k^{f_m})$ is the unknown parameter vector to be identified, $\Delta\theta(k^{f_m})$ is the parameter variation vector, f_m is the mth parametric fault index, and $\delta_u^{\max}$ is the known maximum sustainable variation range of the uth parameter. What we need to pay attention to is that $\Delta\theta_u(k^{f_0})$ is 0 in the normal state and the varied parameter vector remains unchanged for a period of time after the faults occur. Thus, the reset set $X^r(k-1)$ can be obtained by the DE-FD-OEF algorithm, which directionally expands $X(k-1)$ to obtain $X^r(k-1)$ according to whether the optimal volume ellipsoid $E_i^t(k-1)$, $i \in \{1, \ldots, n\}$ containing the test set $X_i^t(k-1)$ is empty.
Step 1: Expand $X(k-1)$ directionally and the test set $X_i^t(k-1)$, $i \in \{1, \ldots, n\}$ can be obtained as follows:

1. when $u \neq i$

$$X_i^t(k-1): \begin{cases} \theta_{x_{i,u}}^{t+}(k-1) = \theta_{x_u}^{+}(k-1) + 2\delta_u^{max} \\ \theta_{x_{i,u}}^{t-}(k-1) = \theta_{x_u}^{-}(k-1) - 2\delta_u^{max}, \end{cases} \tag{7.20}$$

2. when $u = i$

$$X_i^t(k-1): \begin{cases} \theta_{x_{i,u}}^{t+}(k-1) = \theta_{x_u}^{+}(k-1) \\ \theta_{x_{i,u}}^{t-}(k-1) = \theta_{x_u}^{-}(k-1). \end{cases} \tag{7.21}$$

Test set $X_i^t(k-1)$ is a set expanded in $n-1$ directions based on $X(k-1)$, and $X_i^t(k-1)$ is expanded in all directions except in the ith direction. Therefore, we can easily get that there are n test sets for a parameter vector with n parameter components.

Step 2: Calculate the Löwner–John ellipsoid $E_i^t(k-1)$ of $X_i^t(k-1)$ by solving the optimization problem.

A Löwner–John ellipsoid is the minimal volume ellipsoid that contains any given convex set. It is a circumscribed ellipsoid with both uniqueness and existence [2]. The name of Löwner–John ellipsoid comes from two eminent mathematicians, Karel Löwner and Fritz John, who made great contributions to the discovery [3]. In this research, the Löwner–John ellipsoid is used to contain the test set $X_i^t(k-1)$ or the reset set $X^r(k-1)$ to get a minimum volume ellipsoid for the calculation of the OVE algorithm. In this chapter, we should note that the test set and reset set are all orthotopic sets. The calculation method of the Löwner–John ellipsoid containing the test set $X_i^t(k-1)$ is as follows.

$$\min \log(\det(Q_i^{-1})),$$

$$\text{s.t.}$$

$$\| Q_i(V_{x_i^t}^p(k-1)) - \theta_{t_i}^c(k-1) \| \leqslant 1, p = 0, \ldots, 2^n - 1$$

$$Q_i \geqslant 0 \tag{7.22}$$

where $V_{x_i^t}^p(k-1)$ is the pth vertex of $X_i^t(k-1)$, $\theta_{t_i}^c(k-1)$ is the center of $E_i^t(k-1)$, and the shape matrix $P_i^t(k-1) = (Q_i^T Q_i)^{-1}$, $Q_i \in \mathbb{R}^{n \times n}$.

Step 3: Obtain the test measurement set $S^t(k)$ that $S^r(k) = S^t(k) = S(k)$.

Step 4: Set the update length of the test set as L, and use the OVE algorithm to calculate the test ellipsoidal set $E_i^t(k+L)$ at time $k+L$. The value of L is optional and it affects the accuracy of fault isolation.

Step 5: According to the state of $E_i^t(k+L), i \in \{1, \ldots, n\}$, the specific expansion directions of the reset set $X^r(k-1)$ can be decided.

Case 1: If $E_j^t(k+L) = \emptyset$ for $j = 1$ to n, the jth component of the parameter vector in reset set $X^r(k-1)$ meets:

$$\begin{cases} \theta_{x_j}^{r+}(k-1) = \theta_{x_j}^{+}(k-1) + 2\delta_j^{max} \\ \theta_{x_j}^{r-}(k-1) = \theta_{x_j}^{-}(k-1) - 2\delta_j^{max}. \end{cases} \tag{7.23}$$

Case 2: If $E^t_j(k + L) \neq \emptyset$ for $j = 1$ to n, the jth component of the parameter vector in reset set $X^r(k - 1)$ meets:

$$\begin{cases} \theta^{r+}_{x_j}(k - 1) = \theta^+_{x_j}(k - 1) \\ \theta^{r-}_{x_j}(k - 1) = \theta^-_{x_j}(k - 1). \end{cases} \tag{7.24}$$

Step 6: Get the reset set $X^r(k - 1)$:

$$X^r(k - 1) = \left\{ \theta : \begin{bmatrix} \theta^{r-}_{x_1}(k - 1) \\ \vdots \\ \theta^{r-}_{x_n}(k - 1) \end{bmatrix} \leqslant \theta \leqslant \begin{bmatrix} \theta^{r+}_{x_1}(k - 1) \\ \vdots \\ \theta^{r+}_{x_n}(k - 1) \end{bmatrix} \right\}. \tag{7.25}$$

Thus, $E^r(k - 1)$ can be calculated based on $X^r(k - 1)$. Therefore, the reset $E^r(k - 1)$, $X^r(k - 1)$, and $S^r(k)$ satisfy $E^r(k - 1) \cap S^r(k) \neq \emptyset$ and $X^r(k - 1) \cap O(k) \neq \emptyset$, then, by continuing this process, fault identification can be achieved:

$$E(k) = \arg_{E^r}[\min\{\text{vol}(E) : E \supset (E^r(k - 1) \cap S^r(k))\}],$$

$$X(k) = X^r(k - 1) \cap O(k). \tag{7.26}$$

Theorem 7.1 *The jth component of the parameter vector is faulty at the condition that $E^t_j(k + L) = \emptyset$ for $j = 1$ to n, and the jth component of the parameter vector in reset set $X^r(k - 1)$ is expanded as follows:*

$$\begin{cases} \theta^{r+}_{x_j}(k - 1) = \theta^+_{x_j}(k - 1) + 2\delta^{\max}_j \\ \theta^{r-}_{x_j}(k - 1) = \theta^-_{x_j}(k - 1) - 2\delta^{\max}_j, \end{cases} \tag{7.27}$$

Otherwise, the jth component of the parameter vector is characterized as non-faulty at the condition that $E^t_j(k + L) \neq \emptyset$, and the jth component of the parameter vector in reset set $X^r(k - 1)$ remains the same.

$$\begin{cases} \theta^{r+}_{x_j}(k - 1) = \theta^+_{x_j}(k - 1) \\ \theta^{r-}_{x_j}(k - 1) = \theta^-_{x_j}(k - 1). \end{cases} \tag{7.28}$$

Proof If the jth component of the parameter vector in reset set $X^r(k - 1)$ is expanded as Eq. (7.27), then the jth component of the parameter vector has a fault, so the parameter vector $\theta(k^{f_m})$ is satisfied with

$$\begin{cases} \theta_u(k^{f_m}) = \theta_u(k^{f_m} - 1), u \neq j \\ \theta_u(k^{f_m}) = \theta_u(k^{f_m} - 1) + \Delta\theta_u(k^{f_m}), u = j. \end{cases} \tag{7.29}$$

Therefore, when the condition $\theta_{x_{i,j}}^{t-}(k-1) \leqslant \theta_j(k^{f_m}) \leqslant \theta_{x_{i,j}}^{t+}(k-1)$ is satisfied by the test set $X_i^t(k-1), i \in \{1, \ldots, n\}$, it can contain the faulty parameter vector. According to Eqs. (7.20) and (7.21), we can obtain the jth component of a parameter vector in test set $X_i^t(k-1)$ by

$$\begin{cases} \theta_{x_{i,j}}^t(k-1) \in [\theta_{x_j}^-(k-1), \theta_{x_j}^+(k-1)], i = j \\ \theta_{x_{i,j}}^t(k-1) \in [\theta_{x_j}^-(k-1) - 2\delta_j^{max}, \theta_{x_j}^+(k-1) + 2\delta_j^{max}], i \neq j. \end{cases} \tag{7.30}$$

Thus, the Löwner–John ellipsoid $E_i^t(k-1)$ of $X_i^t(k-1)$ satisfies

$$\begin{cases} \theta(k^{f_m}) \notin E_i^t(k-1), i = j \\ \theta(k^{f_m}) \in E_i^t(k-1), i \neq j. \end{cases} \tag{7.31}$$

And whether there is $\theta(k^{f_m})$ in $E_i^t(k+L)$ is consistent with $E_i^t(k-1)$ after L updates. So $E_j^t(k+L)$ is an empty set.

Similarly, we can get that if the jth component of the parameter vector in reset set $X^r(k-1)$ is expanded as Eq. (7.28), then the jth component of the parameter vector is non-faulty, and $E_j^t(k+L)$ is a non-empty set.

Remark 7.1 The case of the directional expansion of a parameter vector with n parameter components is represented in the Theorem 7.1. The directions of faulty parameter vector can be determined according to whether the test sets $X_i^t(k-1)$ and $E_i^t(k-1), i \in \{1, \ldots, n\}$ contain the faulty parameter vector $\theta(k^{f_m})$. ∎

In the following part, we will show the process of the fault diagnosis of systems by using the DE-FD-OEF algorithm mainly from the following three aspects.

7.2.1 Fault Detection

Based on the consistency test of the ellipsoidal set $E(k)$ and the measurement set $S(k)$, the fault detection is implemented. If the formula $E(k) = E(k-1) \cap S(k) = \emptyset$ holds, then there is a fault at time k. The computation time based on the ellipsoidal set is short because of its less computational complexity. Thus, the system's faults can be detected quickly. In Algorithm 7.1, fault indication signal f is used to indicate the fault state of the system and $f = 1$ indicates that the system is detected to be faulty.

7.2.2 Fault Isolation

Fault isolation is achieved according to the state of $E_i^t(k+L), i \in \{1, \ldots, n\}$. n different test sets $X_i^t(k-1), i \in \{1, \ldots, n\}$ are obtained by expanding $X(k-1)$ in $n-1$ directions at the sample time before the system's faults are detected. Then the corresponding Löwner–John ellipsoid $E_i^t(k-1)$ of $X_i^t(k-1)$ is calculated and fault isolation is achieved based on the state of $E_i^t(k+L)$ after L recursive updates. In Theorem 1, an illustrative explanation is given that explains the procedure to isolate the specific faulty components of the parameter vector. Combine the above two cases, the jth component is isolated as faulty if $E_j^t(k+L) = \emptyset$ for $j = 1$ to n, otherwise the jth component is isolated as non-faulty.

7.2.3 Fault Identification

The upper and lower bounds of the feasible parameter set are defined by the intervals $[\theta_{x_u}^+(k), \theta_{x_u}^-(k)]$, $u \in \{1, \ldots, n\}$ of $X(k)$. If the uth component of the parameter vector is isolated as faulty after fault isolation, then the variation and specific value of the uth component can be obtained according to the center of the interval. Simultaneously, we can further obtain the specific value of each parameter component by analyzing the center of $X(k)$ or $E(k)$ at time k. Thus, the information of non-faulty components can be obtained according to these values. The analysis and diagnosis of the whole system are realized through the detailed analysis of the faulty components and non-faulty components.

In summary, for the parameter vector with n parameter components, the GE-FD-OEF algorithm and DE-FD-OEF algorithm can be summarized as Algorithms 7.2 and 7.3.

Remark 7.2 The feasible set of parameter vector does not contain the true value of system's parameter if a fault is detected at time k. It is necessary to expand the range of the set based on $X(k-1)$ in order to further study the system's fault. In most cases, the faulty parameter vector of the system fails in some directions instead of all directions. Thus, the set $X(k-1)$ expanded only in the faulty directions can contain the true value of the parameter. Based on this, the DE-FD-OEF algorithm only expands $X(k-1)$ in faulty directions according to the specific faulty components of the parameter vector, and does not expand in the directions of no fault. Hence, the size of the reset set $X^r(k-1)$ is smaller than that obtained by the GE-FD-OEF algorithm, which expands the $X(k-1)$ in all directions regardless of the faulty directions. Therefore, the corresponding $E^r(k-1)$ is smaller using the DE-FD-OEF algorithm. The set containing the faulty parameter vector can shrink to a certain small range more quickly from set $E^r(k-1)$ than from set $E^{r'}(k-1)$ using the OVE algorithm. Thus, the interval between the upper and lower bounds of X is smaller, the true value of system's fault can be determined faster. The convergence rate of recursive update with OVE algorithm is faster and the fault identification time is shorter. ∎

The recursive update process of DE-FD-OEF and GE-FD-OEF algorithms are shown in Fig. 7.1 when $n = 2$. The red lines and black lines are used to represent the set boundaries of the DE-FD-OEF algorithm and GE-FD-OEF algorithm, respectively. (a) shows that a fault f_m occurred at time k and the parameter vector failed

Algorithm 7.2 GE-FD-OEF algorithm

1: Initialization: $\theta^c(0) = \mathbf{1}_n/p_0$, $P(0) = p_0\mathbf{I}_n$, $\theta_x^+(0) = \theta_o^+(0)$, $\theta_x^-(0) = \theta_o^-(0)$, γ, $\delta^{\max}$, l, $\rho = 10^{-5}$, $p_0 = 10^4$

2: $N \leftarrow$ Constant1

3: **for** $k = 1 : N$ **do**

4:     Obtain input-output data $\{u(k), y(k)\}$ at time instant k

5:     Compute $\theta^c(k)$, $P(k)$, and $f(k)$ for $E(k)$ using *Algorithm 7.1*

6:     **if** $f(k) == 1$ **then**

7:         **for** $u = 1 : n$ **do**

8:             $\theta_{x_u}^{r+}(k-1) \leftarrow \theta_{x_u}^+(k-1) + 2\delta_u^{\max}$

9:             $\theta_{x_u}^{r-}(k-1) \leftarrow \theta_{x_u}^-(k-1) - 2\delta_u^{\max}$

10:         **end for**

11:         **for** $u = 1 : n$ **do**

12:             $\theta_{x_u}^+(k-1) \leftarrow \theta_{x_u}^{r+}(k-1)$

13:             $\theta_{x_u}^-(k-1) \leftarrow \theta_{x_u}^{r-}(k-1)$

14:         **end for**

15:         Calculate $\theta_r^c(k-1)$ and $P^r(k-1)$ for $E^r(k-1)$ referred to Eq. (7.22) based on $X^r(k-1)$

16:         Compute $\theta^c(k)$, $P(k)$, and $f(k)$ for $E(k)$ using *Algorithm 7.1* based on $E^r(k-1)$

17:     **end if**

18:     **for** $u = 1 : n$ **do**

19:         $\theta_{o_u}^+(k) \leftarrow \langle l_u, \theta^c(k)\rangle + \langle l_u, P(k)l_u\rangle^{1/2}$

20:         $\theta_{o_u}^-(k) \leftarrow \langle l_u, \theta^c(k)\rangle - \langle l_u, P(k)l_u\rangle^{1/2}$

21:         $\theta_{x_u}^+(k) \leftarrow \min(\theta_{x_u}^+(k-1), \theta_{o_u}^+(k))$

22:         $\theta_{x_u}^-(k) \leftarrow \max(\theta_{x_u}^-(k-1), \theta_{o_u}^-(k))$

23:         $\theta_{x_u}^c(k) \leftarrow \frac{\theta_{x_u}^+(k)+\theta_{x_u}^-(k)}{2}$

24:     **end for**

25: **end for**

---

in the θ_2-axis direction, changing from $\theta(k^{f_m} - 1)$ to $\theta(k^{f_m})$. To contain the varied parameter vector, two algorithms are used to expand the feasible parameter set on the basis of the original set $X(k - 1)$. The DE-FD-OEF algorithm only expands in the faulty direction, and the reset set $X^r(k - 1)$ of the DE-FD-OEF algorithm is considerably smaller than the reset set $X^{r'}(k - 1)$ obtained by the GE-FD-OEF algorithm. Accordingly, the Löwner–John ellipsoid $E^r(k - 1)$ of $X^r(k - 1)$ is considerably smaller than that of $X^{r'}(k - 1)$. (b) and (c) show the recursive update process based on the reset ellipsoidal sets $E^r(k - 1)$ and $E^{r'}(k - 1)$ after the expansion. It can be seen clearly that the update ellipsoidal sets corresponding to the DE-FD-OEF algorithm are smaller; therefore, the DE-FD-OEF algorithm can converge to the true value of the parameter vector at a faster pace than the GE-FD-OEF algorithm.

Algorithm 7.3 DE-FD-OEF algorithm

1: Initialization: $\theta^c(0) = 1_n/p_0$, $P(0) = p_0 I_n$, $\theta_x^+(0) = \theta_o^+(0)$, $\theta_x^-(0) = \theta_o^-(0)$, γ, $\delta^{\max}$, l, $\rho = 10^{-5}$, $p_0 = 10^4$

2: $N \leftarrow$ Constant1

3: Update length of test set $L \leftarrow$ Constant2

4: **for** $k = 1 : N$ **do**

5:     Obtain input-output data $\{u(k), y(k)\}$ at time instant k

6:     Compute $\theta^c(k)$, $P(k)$, and $f(k)$ for $E(k)$ using *Algorithm 7.1*

7:     **if** $f(k) == 1$ **then**

8:         Obtain $X_i^t(k-1)$, $i = 1, \ldots, n$ using Eq. (7.20) and (7.21)

9:         Calculate $E_i^t(k-1)$, $i = 1, \ldots, n$ using Eq. (7.22)

10:        **for** $g = k : k + L$ **do**

11:            Compute $\theta_{t_i}^c(g)$, $P_i^t(g)$ and $f_i^t(g)$ for $E_i^t(g)$, $i = 1, \ldots, n$ using *Algorithm 7.1*

12:        **end for**

13:        **for** $i = 1 : n$ **do**

14:            **if** $f_i^t(k+L) == 1$ **then**

15:                $\theta_{x_i}^{r+}(k-1) \leftarrow \theta_{x_i}^+(k-1) + 2\delta_i^{\max}$

16:                $\theta_{x_i}^{r-}(k-1) \leftarrow \theta_{x_i}^-(k-1) - 2\delta_i^{\max}$

17:            **else**

18:                $\theta_{x_i}^{r+}(k-1) \leftarrow \theta_{x_i}^+(k-1)$

19:                $\theta_{x_i}^{r-}(k-1) \leftarrow \theta_{x_i}^-(k-1)$

20:            **end if**

21:        **end for**

22:        **for** $u = 1 : n$ **do**

23:            $\theta_{x_u}^+(k-1) \leftarrow \theta_{x_u}^{r+}(k-1)$

24:            $\theta_{x_u}^-(k-1) \leftarrow \theta_{x_u}^{r-}(k-1)$

25:        **end for**

26:        Calculate $\theta_r^c(k-1)$ and $P^r(k-1)$ for $E^r(k-1)$ referred to Eq. (7.22) based on $X^r(k-1)$

27:        Compute $\theta^c(k)$, $P(k)$, and $f(k)$ for $E(k)$ using *Algorithm 7.1* based on $E^r(k-1)$

28:    **end if**

29:    **for** $u = 1 : n$ **do**

30:        $\theta_{o_u}^+(k) \leftarrow \langle l_u, \theta^c(k) \rangle + \langle l_u, P(k)l_u \rangle^{1/2}$

31:        $\theta_{o_u}^-(k) \leftarrow \langle l_u, \theta^c(k) \rangle - \langle l_u, P(k)l_u \rangle^{1/2}$

32:        $\theta_{x_u}^+(k) \leftarrow \min(\theta_{x_u}^+(k-1), \theta_{o_u}^+(k))$

33:        $\theta_{x_u}^-(k) \leftarrow \max(\theta_{x_u}^-(k-1), \theta_{o_u}^-(k))$

34:        $\theta_{x_u}^c(k) \leftarrow \frac{\theta_{x_u}^+(k) + \theta_{x_u}^-(k)}{2}$

35:    **end for**

36: **end for**

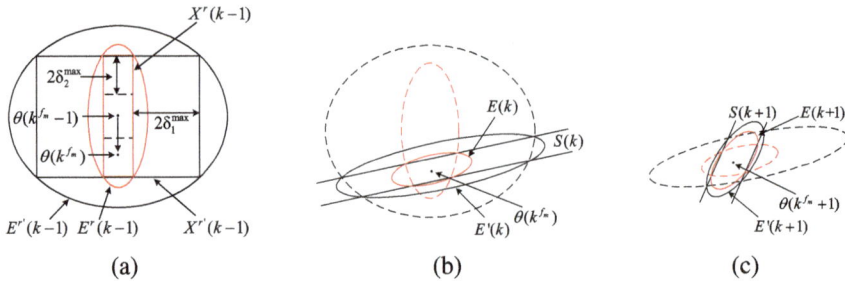

Fig. 7.1 Recursive process of DE-FD-OEF and GE-FD-OEF algorithms

7.3 Orthotopic Double Filtering Based State Estimation Algorithm

7.3.1 Prediction Step

Suppose that at step k, the feasible set of states is a orthotope $\mathcal{O}(\overline{x}(k), \overline{d}(k))$, at the same time $Z(\overline{x}(k), \overline{D}(k)) = \mathcal{O}(\overline{x}(k), \overline{d}(k))$, $\overline{D}(k)$ is a diagonal matrix whose diagonal value is equal to d. Without considering system disturbances,

$$\overline{x}(k + 1|k) = A\overline{x}(k) + Bu(k), \qquad (7.32)$$

$$D(k + 1|k) = A\overline{D}(k). \qquad (7.33)$$

It can be seen that when A is not a diagonal matrix, then the feasible set obtained by the prediction step is no longer a orthotope, but a zonotope of arbitrary shape. Convert this zonotope into the most compact orthotope. At this time, $D(k + 1|k)$ is a diagonal matrix. When considering the system disturbance W, according to the Minkowski sum of orthotopic bodies, because $W = \mathcal{O}(0, w)$, then $\overline{x}(k + 1|k)$ remains unchanged,

$$\overline{D}(k + 1|k) = D(k + 1|k) \oplus w. \qquad (7.34)$$

7.3.2 Update Step

Knowing that the feasible set of the predicted state obtained by the prediction step is a orthotope, it is easy to convert it into $2n_x$ constraints and express them as $\overline{(X)}(k)$.

Suppose there is a fully symmetrical polysome $Z(p, H) = p \oplus H\mathbf{B}^n$, the system state is satisfied at a certain moment $\mathcal{S} = \{x \in \mathbb{R}^n : |c^{\mathrm{T}}x - d| \leq \sigma\}$, Then $X \cap \mathcal{S} \subseteq \hat{Z}(\hat{p}, \hat{H}) = \hat{P}(\lambda) \oplus \hat{H}(\lambda)\mathbf{B}^{r+1}$. Where

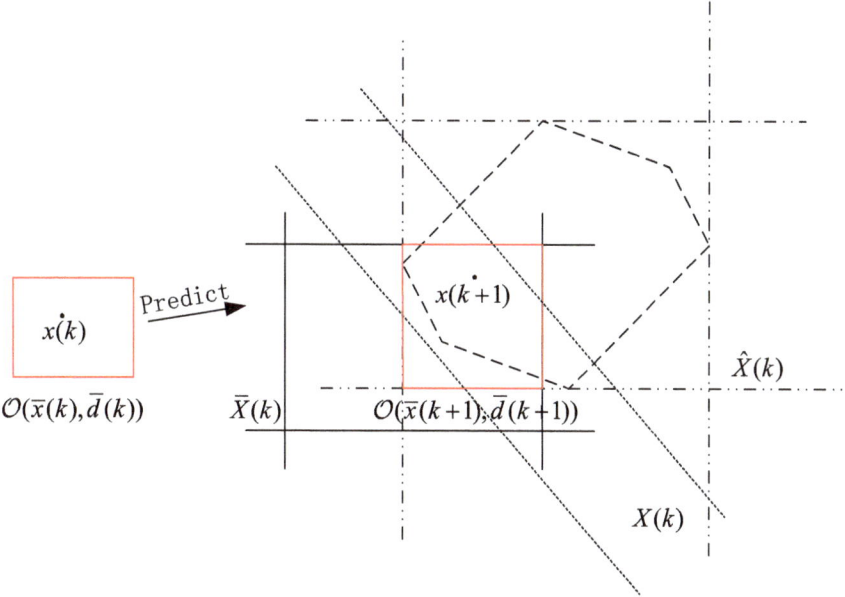

Fig. 7.2 State estimation graph based on orthotopic double filtering

$$\hat{p}(\lambda) = p + \lambda \left(d - c^{\mathrm{T}} p\right), \tag{7.35}$$

$$\hat{H}(\lambda) = \left[\left(I - \lambda c^{\mathrm{T}}\right) H \quad \sigma \lambda\right]. \tag{7.36}$$

When selecting the minimum edge criterion, $\lambda^* = \frac{HH^{\mathrm{T}}c}{c^{\mathrm{T}}HH^{\mathrm{T}}c+\sigma^2}$.

Next, wrapping the above-mentioned zonotope $Z(p, H)$ with the most compact orthotope. At this time, the upper and lower bounds of the zonotope have not changed. After wrapping the compact orthotopes and discretizing them into $2n_x$ constraints and denoting them as $\hat{X}(k)$, by solving $2n_x$ linear programming problems, the upper and lower bounds of each state are obtained. The specific principle is shown in Fig. 7.2.

Using linear programming to solve the upper and lower bounds of the current orthotope:

$$\beta_i(k + 1) = \max e_i^{\mathrm{T}} x,$$
$$\text{s.t.}$$
$$x \in X(k) \cap \bar{X}(k) \cap \hat{X}(k). \tag{7.37}$$

$$\beta_{i+n_x}(k + 1) = \min e_i^{\mathrm{T}} x,$$
$$\text{s.t.}$$
$$x \in X(k) \cap \bar{X}(k) \cap \hat{X}(k). \tag{7.38}$$

Table 7.1 Specific values of parameter components at different states

State	k	θ_1	θ_2	θ_3
1	[1, 1000]	−1.0000	1.0000	1.0000
2	[1001, 2000]	−1.5000	1.0000	0.5000
3	[2001, 3000]	−1.5000	1.0000	1.0000
4	[3001, 4000]	−1.5000	0.7000	1.0000

$$X(k+1) = \bigcap_{i=1}^{2n_x} (X_i(k) \cap \bar{X}_i(k) \cap \hat{X}_i(k)). \tag{7.39}$$

Therefore, the feasible set of the state at time $k+1$ is orthotope $\mathcal{O}(\bar{x}(k+1), \bar{d}(k+1))$, $i = 1, 2, \ldots, n_x$, where

$$\bar{x}(k+1)^i = \frac{\beta_i^i(k+1) + \beta_{i+n_x}^i(k+1)}{2}, \tag{7.40}$$

$$\bar{d}(k+1)^i = \frac{\beta_i^i(k+1) - \beta_{i+n_x}^i(k+1)}{2}; \tag{7.41}$$

7.4 Illustrative Simulation

Consider a linear discrete-time system as follow

$$y(k) = \phi^{\mathrm{T}}(k)\theta + e(k) = \begin{bmatrix} -y(k-1), & -y(k-2), & u(k) \end{bmatrix} \begin{bmatrix} \theta_1 \\ \theta_2 \\ \theta_3 \end{bmatrix} + e(k) \tag{7.42}$$

with $k = 1, 2, \ldots, N$, $|e(k)| \leqslant 0.1$, and $u(k) \in U[-1, 1]$. It is note that the total number of parameter components is $n = 3$ and the system is assumed to have four states, including one normal state and three faulty states. The specific values of parameter components at different states are given by Table 7.1. The system is healthy in the initial period, and the fault 1, fault 2, and fault 3 are added when $k \in [1001, 2000]$, $k \in [2001, 3000]$, and $k \in [3001, 4000]$, respectively, i.e., the system is in the second state, third state, and fourth state, respectively.

In this chapter, the fault diagnosis of this system is implemented by the proposed DE-FD-OEF algorithm and GE-FD-OEF algorithm, and Figs. 7.3, 7.4 and 7.5 show the upper and lower bounds of the orthotopic set $X(k)$. The final estimation values of the parameter components for each state are shown in Table 7.1.

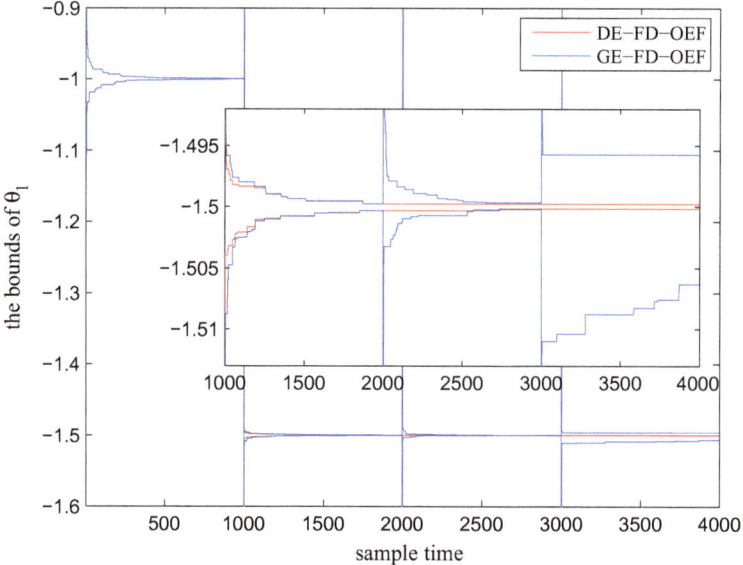

Fig. 7.3 Upper ($\theta_{x_1}^+$) and lower ($\theta_{x_1}^-$) bounds of θ_1 via two algorithms

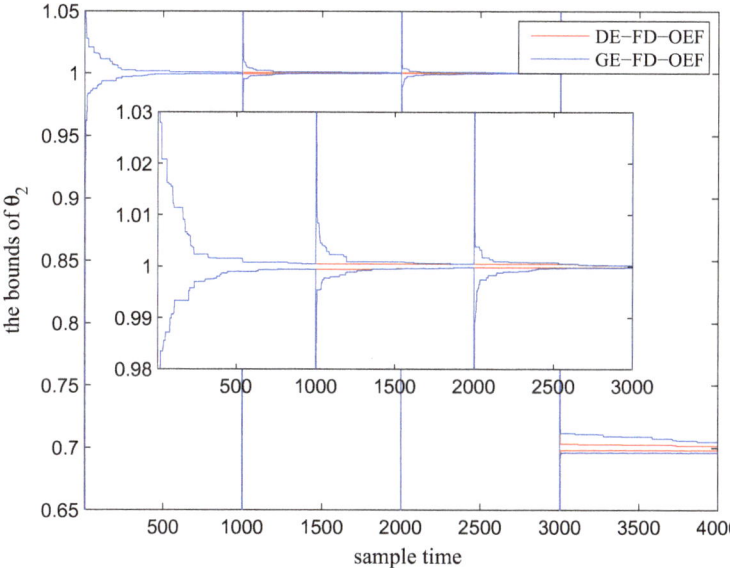

Fig. 7.4 Upper ($\theta_{x_2}^+$) and lower ($\theta_{x_2}^-$) bounds of θ_2 via two algorithms

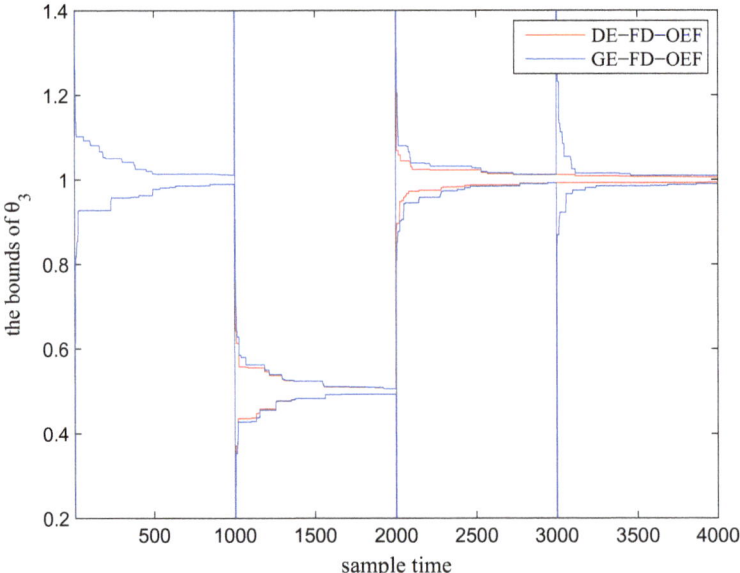

Fig. 7.5 Upper $(\theta_{x_3}^+)$ and lower $(\theta_{x_3}^-)$ bounds of θ_3 via two algorithms

From Figs. 7.3, 7.4 and 7.5, it can be concluded that:

1. There are three reset times in the upper and lower bounds of $X(k)$ based on the DE-FD-OEF and GE-FD-OEF algorithms, indicating the system has four states, containing a normal state and three faulty states.

2. The first state is considered a normal state, in which the upper and lower bounds of $X(k)$ obtained by the two algorithms are coincident. Table 7.2 lists the final estimation values of the parameter components of each state under the two algorithms. It should be noted that θ_x^c is the center of $X(k)$, which can be used to represent the estimation result of the parameter vector. Here, θ_x^c and $\theta_x^{c'}$ are the final estimation values of the parameter vector obtained by the DE-FD-OEF and GE-FD-OEF algorithms, respectively. From Table 7.2, the two algorithms have the same estimation results in this normal state, and they are all very close to the true values of the parameter vector. Thus, we can conclude that the DE-FD-OEF and GE-FD-OEF algorithms yield equally high accuracy during parameter estimation.

3. The upper and lower bounds of $X(k)$ of the two algorithms are reset at $k = 1001$, $k = 2001$, and $k = 3001$, which indicates that the system has faults at these three moments. Thus, the speed of fault detection is consistent on using both the DE-FD-OEF and GE-FD-OEF algorithms and they can detect the system's faults in time.

4. After fault detection, the reset of each component of the parameter vector is different. For the DE-FD-OEF algorithm, it will isolate the faults to obtain specific faulty components after fault detection before dealing with the faults. Without loss of generality, state 2 is taken as an example for simulation analysis and the

Table 7.2 Final estimation values of parameter components for each state

k	$\theta_{x_1}^c$	$\theta_{x_2}^c$	$\theta_{x_3}^c$	$\theta_{x_1}^{c\,\prime}$	$\theta_{x_2}^{c\,\prime}$	$\theta_{x_3}^{c\,\prime}$
1000	−1.0000	0.9999	0.9995	−1.0000	0.9999	0.9995
2000	−1.5000	1.0001	0.4983	−1.5000	1.0001	0.4983
3000	−1.4999	1.0000	1.0025	−1.4999	1.0000	1.0031
4000	−1.4999	0.6998	1.0008	−1.5003	0.7003	1.0003

analyses of state 3 and state 4 can refer to this analysis process. As we can see, only the upper and lower bounds of θ_1 and θ_3 are reset and re-converged in state 2, while the bounds of θ_2 keep the original convergence trend by using DE-FD-OEF algorithm. However, for the GE-FD-OEF algorithm, the upper and lower bounds of θ_1, θ_2 and θ_3 are all reset and re-converged. Thus, we can isolate the first and third components as faulty using the DE-FD-OEF algorithm.

5. From the comparison of the upper and lower bounds based on the two algorithms, we can get that the upper and lower bounds of θ_2 are not reset by the DE-FD-OEF algorithm, whose intervals between the upper and lower bounds are significantly smaller than that of the GE-FD-OEF algorithm, other components of the parameter vector also have intervals with a faster convergence rate. Therefore, the DE-FD-OEF algorithm has a faster convergence rate during fault diagnosis than the GE-FD-OEF algorithm.

6. In the early stage of each state, the intervals of the bounds of the GE-FD-OEF algorithm are much larger than those of the DE-FD-OEF algorithm. However, as the recursive time increases, the interval bounds gradually become closer, and both have similar estimation results eventually. Based on Table 7.2, both algorithms can obtain accurate estimation values of faulty components. Thus, the DE-FD-OEF and GE-FD-OEF algorithms all have high accuracy of fault identification.

In addition, Fig. 7.6 shows the estimation errors of the parameter vector of the two algorithms. The estimation error of the parameter vector is defined as

$$\delta_{\theta_x^c} = \frac{\|\,\hat{\theta}_x^c(k) - \hat{\theta}_{x_o}^c\,\|}{\|\,\hat{\theta}_{x_o}^c\,\|} \times 100\%. \tag{7.43}$$

Here, $\hat{\theta}_{x_o}^c$ is the final estimation value of the parameter vector for each state. Comparing the two curves, the curve obtained by the DE-FD-OEF algorithm drops faster and approaches 0 faster than the GE-FD-OEF algorithm. Furthermore, the fault identification time is defined according to the estimation error of the parameter vector that is smaller than a certain threshold value within a certain time length:

$$\frac{\|\,\hat{\theta}_x^c(k+h) - \hat{\theta}_{x_o}^c\,\|}{\|\,\hat{\theta}_{x_o}^c\,\|} \times 100\% \leqslant \varepsilon, \ h = 1, 2, \ldots, H. \tag{7.44}$$

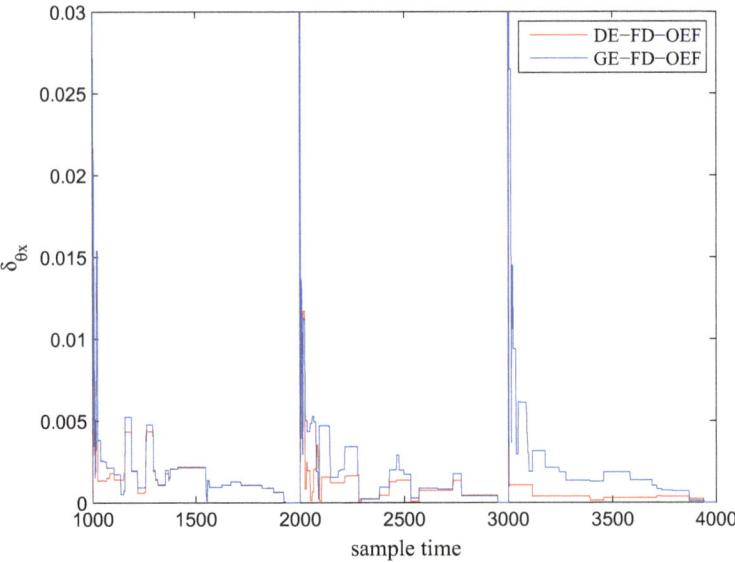

Fig. 7.6 Estimation errors of θ via two algorithms

H and ε represent the selected time length and the given threshold value, respectively. If the estimation error of the parameter vector is smaller than ε during $k + 1$ to $k + H$, then we can consider that the algorithm has identified the faults of the system at time $k + 1$ accurately. Set $H = 150$ and $\varepsilon = 0.5$ in the simulation of this chapter. Then the time $k + 1$ satisfying the above inequality in the two algorithms are 1026 and 1187, respectively, which shows that the DE-FD-OEF algorithm has a faster fault identification speed. Based on the analysis of fault state 1, fault state 2 and 3 can be analyzed in a similar manner, and similar conclusions can be drawn.

In Fig. 7.7, the partial recursive evolution process of $X(k)$ obtained by the DE-FD-OEF (red lines) and GE-FD-OEF (blue lines) algorithms in fault state 1 are selected. It can be seen from Fig. 7.7 that the volume of $X(k)$ obtained by the DE-FD-OEF algorithm is smaller than that of GE-FD-OEF algorithm. The interval of $X(k)$ on the θ_2-axis obtained by the DE-FD-OEF algorithm is very small, which implies that no expansion is made in this direction. Meanwhile, the intervals in the other two directions are also smaller than those of the GE-FD-OEF algorithm. To the best of our knowledge, the smaller the volume of the orthotope bounding the feasible parameter set, the closer the estimation value of the parameter vector is to the true value of parameter vector, and the faster the fault identification can be realized. Therefore, the GE-FD-OEF algorithm is conservative compared to the DE-FD-OEF algorithm and features a slower convergence rate.

In order to make comparison of the DE-FD-OEF algorithm with the GE-FD-OEF algorithm, the orthotopic set $X(k)$ of three-dimension is further projected in the two-dimensional plane, and its shape is shown as a rectangle. The corresponding

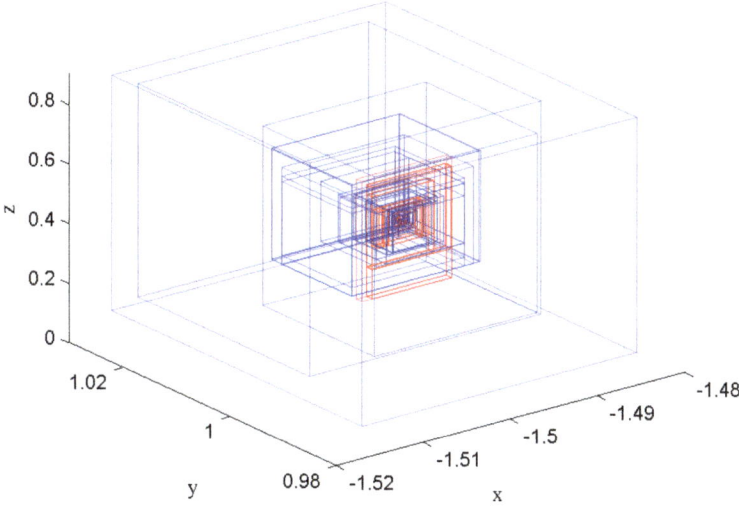

Fig. 7.7   Recursive evolution process of $X(k)$ based on DE-FD-OEF and GE-FD-OEF algorithms

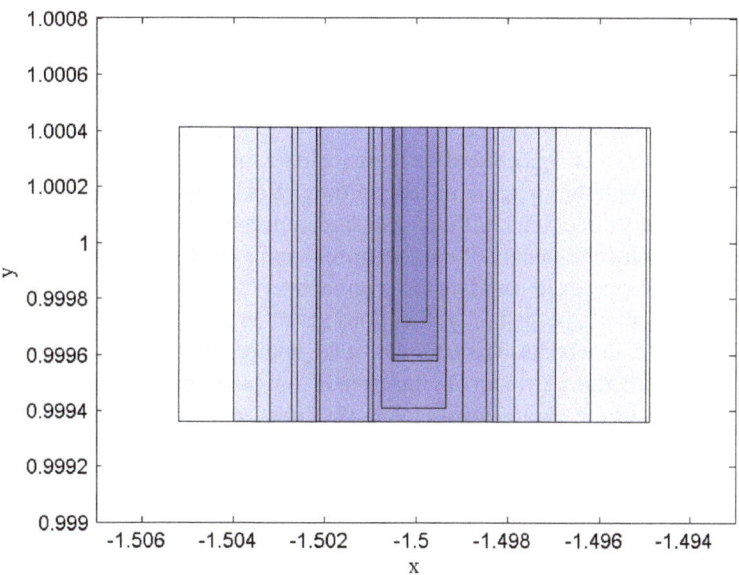

Fig. 7.8   Recursive evolution process of $X(k)$ in two-dimensional plane based on DE-FD-OEF algorithm

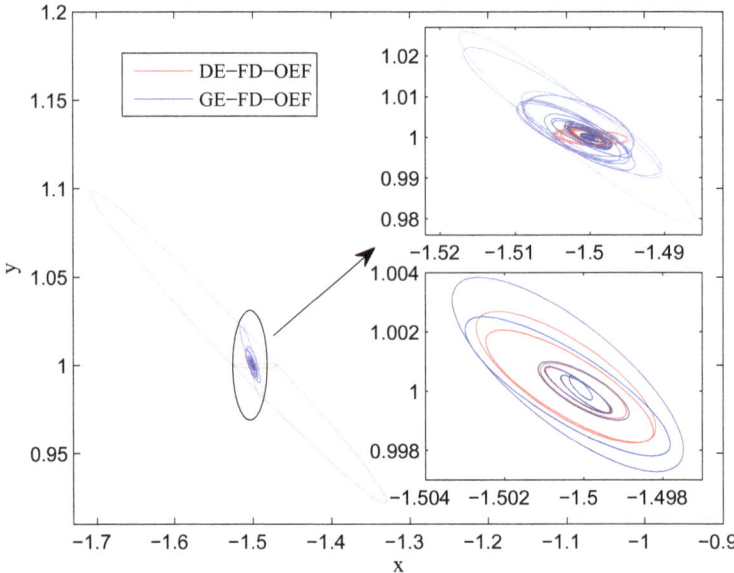

Fig. 7.9 Recursive evolution process of ellipsoidal set $E(k)$ in two-dimensional plane based on DE-FD-OEF and GE-FD-OEF algorithms

projection is shown as Fig. 7.8, and the increase in sample time k is shown by the depth of blue color. With the increase in sample time, $X(k)$ has an obvious convergence trend in the faulty direction, i.e., the θ_1-axis direction, and gradually shrinks to the true value of the faulty component of the parameter vector. In the θ_2-axis direction, $X(k)$ has a further convergence based on the original interval. Therefore, the DE-FD-OEF algorithm only resets in the faulty directions based on the accurate directions of the specific faulty components. The intervals of the components of the parameter vector in the faulty directions have an obvious trend from large to small, and the intervals of components of the parameter vector in the fault-free directions further converge based on the original intervals. At this time, the range of convergence is limited based on the small original intervals because of the conservatism of the orthotopic set obtained by the DE-FD-OEF algorithm.

Figures 7.9 and 7.10 analyze the recursive evolution process of the two algorithms in fault state 1 from the perspective of an ellipsoid. Similarly, for the convenience of demonstration, the ellipsoidal set $E(k)$ is projected on the two-dimensional plane to obtain the corresponding ellipse.

In Fig. 7.9, the process which causes the area of the ellipse to decreases with the increase in sample time is mainly focused. As we can see in this process, the area of ellipse based on the DE-FD-OEF algorithm is obviously smaller than that of GE-FD-OEF algorithm in the early period of fault state 1, which shows that DE-FD-OEF algorithm can shrink to the true value of the parameter vector faster and has the advantage of a fast convergence rate. As the ellipsoidal set of the set membership

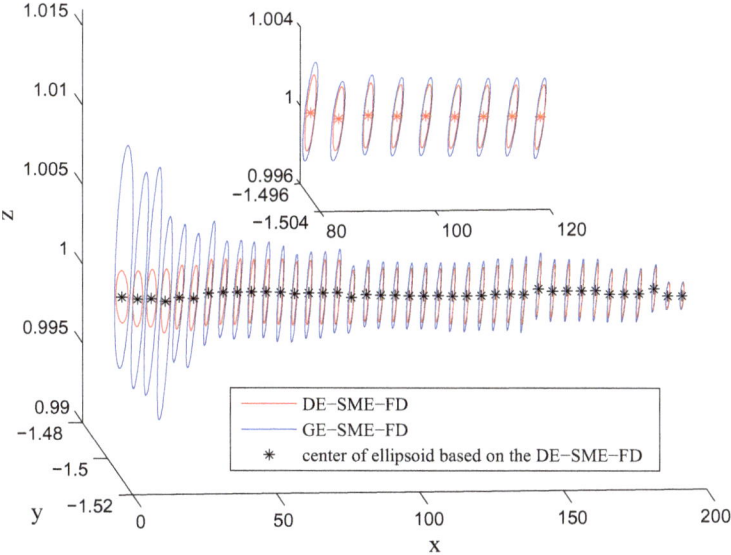

Fig. 7.10 Recursive evolution process of ellipsoidal set $E(k)$ in two-dimensional plane based on DE-FD-OEF and GE-FD-OEF algorithms and center of ellipsoid based on DE-FD-OEF algorithm

estimation algorithm has a certain convergence limitation with the increase in time, the final areas of the ellipse and final estimation values of the parameter vector obtained by the two algorithms are similar. It should be noted that the change of the ellipse with time is represented by the color depth of the elliptical boundary.

From Fig. 7.10, we can see the recursive evolution process of ellipse in the early period of fault state 1 based on the two algorithms and the estimation center of ellipse based on the DE-FD-OEF algorithm. And the ellipse based on the two algorithms gradually narrows in the θ_1-axis direction is show in Fig. 7.10. In the θ_2-axis direction, the GE-FD-OEF algorithm has an obvious convergence trend, while the DE-FD-OEF algorithm is roughly unchanged at the beginning, and then has a small convergence. The results presented by these descriptions are consistent with the analysis of the set $X(k)$ described above.

7.5 Application Study

7.5.1 Application Case 1

The spring-damping system shown in Fig. 7.11 is simulated in order to further verify the proposed fault diagnosis algorithm. The spring-damping system has a wide range of applications in daily life, and its mathematical model can be expressed as [4]:

Fig. 7.11 Spring-damping system

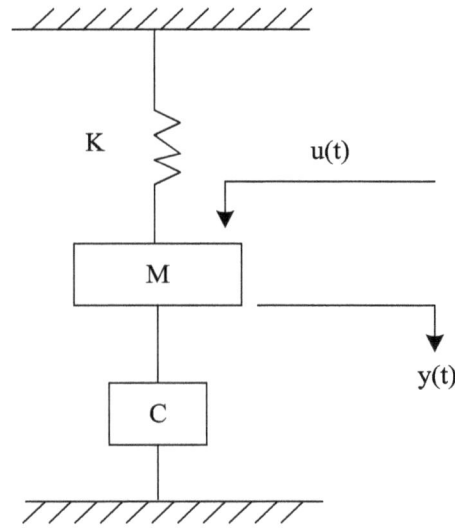

$$\begin{cases} \dot{x}_1(t) = x_2(t), \\ \dot{x}_2(t) = -\dfrac{K}{M}x_1(t) - \dfrac{C}{M}x_2(t) + \dfrac{1}{M}u(t). \end{cases} \tag{7.45}$$

Where $M = 1\,\text{kg}, K = 3\,\text{N/m}, C = 0.5\,\text{Ns/m}, u(t) \in U[-15, 15]$ define the mass, spring constant, damping coefficient and external control force, respectively. Let $y(t) = x_1(t), \dot{y}(t) = x_2(t)$, then the above system model can be written as

$$\frac{y(s)}{u(s)} = \frac{\frac{1}{M}}{s^2 + \frac{C}{M}s + \frac{K}{M}}. \tag{7.46}$$

The sampling time T_s is set as $0.1s$ to discretize the second-order system and (7.46) can be written as

$$\begin{aligned} y(k) &= \phi^{\mathrm{T}}\theta + e(k) \\ &= 1.9220y(k-1) - 0.9512y(k-2) \\ &\quad + 0.0049u(k-1) + 0.0048u(k-2) + e(k) \end{aligned} \tag{7.47}$$

with $k = 1, 2, \ldots, N$, $e(k) \in U[-0.01, 0.01]$ is the unknown but bounded noise, $\theta = [\theta_1, \theta_2, \theta_3, \theta_4]^{\mathrm{T}}$ is the parameter vector.

Fault 1, fault 2, and fault 3 are added as Table 7.3 to the system when k meets 1001, 2001, and 3001, respectively. Then the fault diagnosis of the system is performed using the algorithm proposed in this chapter. The fault diagnosis results are shown in Figs. 7.12, 7.13, 7.14 and 7.15 and Table 7.4.

Table 7.3 Parameter components of the spring-damping system at different states

State	k	θ_1	θ_2	θ_3	θ_4
1	[1, 1000]	−1.9220	0.9512	0.0049	0.0048
2	[1001, 2000]	−1.9220	0.9512	0.0074	0.0072
3	[2001, 3000]	−1.7620	0.9512	0.0045	0.0072
4	[3001, 4000]	−1.7620	0.7860	0.0023	0.0052

The recursive evolution process of the upper and lower bounds of the feasible parameter set $X(k)$ are used to demonstrate the fault diagnosis results of the system. From Figs. 7.12, 7.13, 7.14 and 7.15 and Table 7.4, we can conclude that:

1. The upper and lower bounds of $X(k)$ have been reset three times of the two algorithms during the whole diagnosis process, indicating that three faults have occurred in this system.
2. In the first state, the two algorithms have the same upper and lower bounds of $X(k)$. And it can be seen from Table 7.4, the estimation values of the parameter components are also the same, which indicates that the first state is the normal state and both the DE-FD-OEF and GE-FD-OEF algorithms achieve parameter estimation with high accuracy.
3. Specific analysis shows that the upper and lower bounds of the parameter components are reset at $k = 1001, 2001$, and 3001, respectively, which implies that the two algorithms can detect the system's faults in time.
4. The DE-FD-OEF algorithm resets θ_{x_3} and θ_{x_4} in fault state 1, resets θ_{x_1} and θ_{x_3} in fault state 2, and resets θ_{x_2}, θ_{x_3} and θ_{x_4} in fault state 3. Thus, we can observe the third and fourth components are isolated as faulty in fault state 1, the first and third components are isolated as faulty in fault state 2, and the second, the third and fourth components are isolated as faulty in fault state 3; these results are consistent with the real fault state of the spring-damping system. Therefore, this algorithm can isolate the system's faults accurately.
5. From the results of the fault isolation, we can conclude that in the DE-FD-OEF algorithm, the components are reset directionally, while in the GE-FD-OEF algorithm, all the components of the parameter vector are reset, leading to the problem of a larger interval between the upper and lower bounds. Consequently, the convergence rate of fault identification in the spring-damping system of GE-FD-OEF algorithm is slower than that of the DE-FD-OEF algorithm.
6. Table 7.4 shows that the final estimation values of parameter components of the two algorithms are very similar, indicating that the DE-FD-OEF and GE-FD-OEF algorithms can identify the fault values of the spring-damping system accurately.

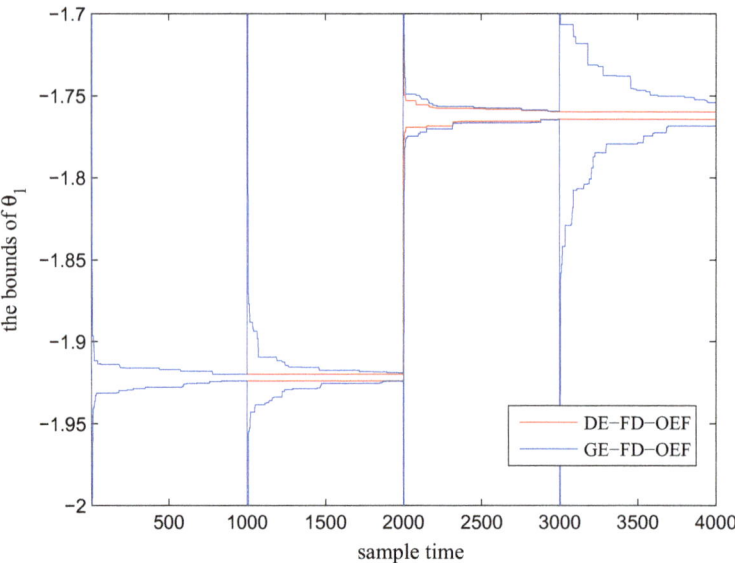

Fig. 7.12 Upper $(\theta_{x_1}^+)$ and lower $(\theta_{x_1}^-)$ bounds of θ_1 via different algorithms

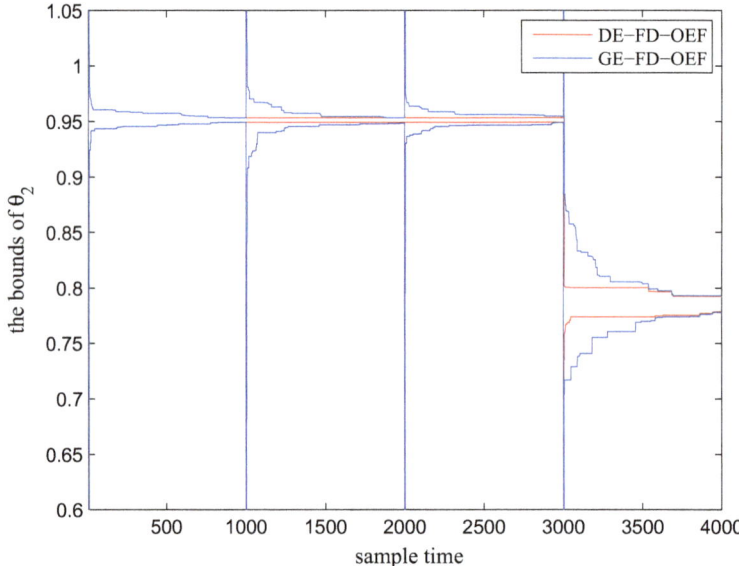

Fig. 7.13 Upper $(\theta_{x_2}^+)$ and lower $(\theta_{x_2}^-)$ bounds of θ_2 via different algorithms

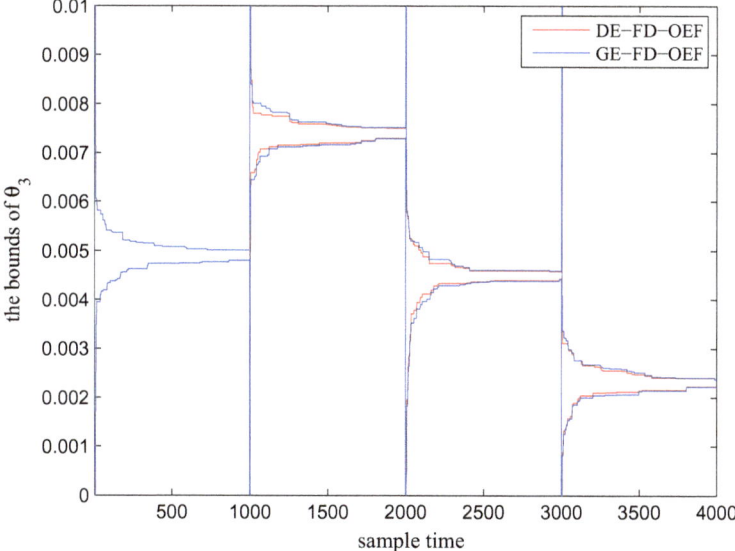

Fig. 7.14 Upper $(\theta_{x_3}^+)$ and lower $(\theta_{x_3}^-)$ bounds of θ_3 via different algorithms

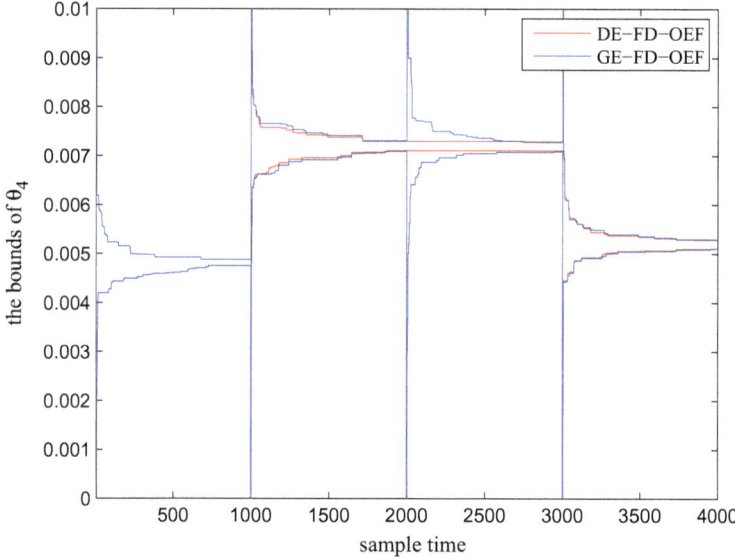

Fig. 7.15 Upper $(\theta_{x_4}^+)$ and lower $(\theta_{x_4}^-)$ bounds of θ_4 via different algorithms

Table 7.4 Final estimation values of parameter components for each state

k	$\theta_{x_1}^c$	$\theta_{x_2}^c$	$\theta_{x_3}^c$	$\theta_{x_4}^c$	$\theta_{x_1}^{c\,\prime}$	$\theta_{x_2}^{c\,\prime}$	$\theta_{x_3}^{c\,\prime}$	$\theta_{x_4}^{c\,\prime}$
1000	−1.9217	0.9510	0.0049	0.0048	−1.9217	0.9510	0.0049	0.0048
2000	−1.9217	0.9508	0.0074	0.0072	−1.9217	0.9509	0.0074	0.0072
3000	−1.7619	0.9514	0.0045	0.0072	−1.7618	0.9513	0.0045	0.0072
4000	−1.7637	0.7876	0.0023	0.0052	−1.7638	0.7878	0.0023	0.0052

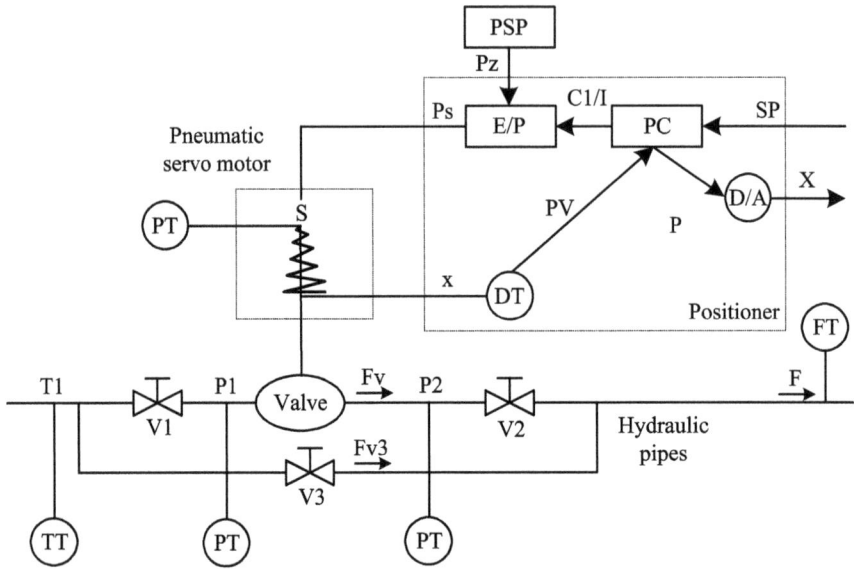

Fig. 7.16 DAMADICS smart actuator block diagram

7.5.2 Application Case 2

As shown in Fig. 7.16, the development and application of methods for actuator diagnosis in industrial control systems (DAMADICS) smart actuator block diagram is discussed in this simulation, and the non-linear second-order dynamics of pneumatic servomotor is described by [5]:

$$m\frac{\mathrm{d}^2 X}{\mathrm{d}t^2} = -k_v\frac{\mathrm{d}X}{\mathrm{d}t} - k_x(k + X) + A_e P_s + mg. \tag{7.48}$$

where m is the mass rod, X is the servomotor rod displacement, A_e is the diaphragm area, P_s is the pressure in the servomotor chamber, k_v is the valve constant, k_x is the spring and diaphragm constant and k is a constant (0.00925).

The model can be replaced by a linearized version after simplifications and discretization:

Table 7.5 Parameter components of the pneumatic servomotor at different states

State	k	θ_1	θ_2	θ_3	θ_4	θ_5
1	[1, 1000]	0.0501	−0.0032	−0.8545	−0.6631	0.5434
2	[1001, 2000]	0.0501	−0.0070	−0.8545	−0.6631	0.5300
3	[2001, 3000]	0.0583	−0.0070	−0.8450	−0.6680	0.5412
4	[3001, 4000]	0.0550	−0.0070	−0.8506	−0.6640	0.5300

$$X(k) = \frac{b_{x_2}q^{-2} + b_{x_3}q^{-3}}{1 + a_{x_1}q^{-1} + a_{x_2}q^{-2} + a_{x_3}q^{-3}} CVP(k). \tag{7.49}$$

Here, $CVP(k)$ is the command pressure. Let $y(k) = X(k), u(k) = CVP(k)$, then the above system model can be replaced by:

$$\begin{aligned}
y(k) &= \frac{b_{x_2}q^{-2} + b_{x_3}q^{-3}}{1 + a_{x_1}q^{-1} + a_{x_2}q^{-2} + a_{x_3}q^{-3}} u(k) \\
&= -a_{x_1}y(k-1) - a_{x_2}y(k-2) - a_{x_3}y(k-3) \\
&\quad + b_{x_2}u(k-2) + b_{x_3}u(k-3) \\
&= \phi^T\theta,
\end{aligned} \tag{7.50}$$

where $\theta = [\theta_1, \theta_2, \theta_3, \theta_4, \theta_5]^T = [a_{x_1}, a_{x_2}, a_{x_3}, b_{x_2}, b_{x_3}]^T$ is the parameter vector, $\phi = [-y(k-1), -y(k-2), -y(k-3), u(k-2), u(k-3)]^T$ is the regression vector. Add the unknown but bounded noise $e(k)$ ($e(k) \in U[-0.001, 0.001]$) to the system, the system's model can be expressed by:

$$y(k) = \phi^T\theta + e(k). \tag{7.51}$$

When k meets $1001, 2001, 3001$, we added fault 1, fault 2 and fault 3 to the system respectively, and specific values of parameter vector θ in different time periods are shown as Table 7.5.

Similarly, the recursive evolution process of upper and lower bounds of the feasible parameter set $X(k)$ are used to demonstrate the fault diagnosis results of the pneumatic servomotor. The upper and lower bounds of θ_1 to θ_5 are shown in Figs. 7.17, 7.18, 7.19, 7.20 and 7.21 respectively, and from these figures, we can draw some conclusions similar to the above examples.

The following three points are simply summarized in order to avoid repetition:

1. Fault detection can be quickly realized by the two algorithms, the system has three different fault states.
2. According to the curves of the upper and lower bounds of the parameter vector obtained by DE-FD-OEF algorithm, we can get that the second and fifth components are faulty in fault state 1, the first, third, fourth and fifth components are

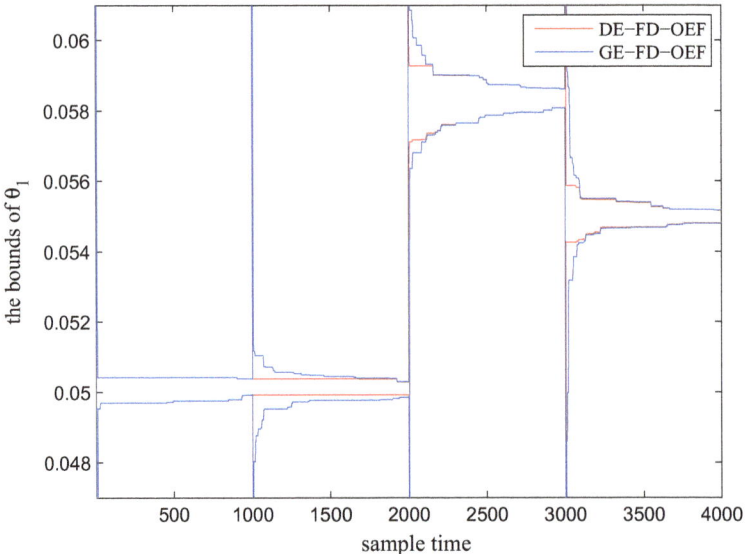

Fig. 7.17 Upper ($\theta_{x_1}^{+}$) and lower ($\theta_{x_1}^{-}$) bounds of θ_1 via different algorithms

faulty in fault state 2, the first, third, fourth and fifth components are faulty in fault state 3, and the system's fault isolation is successfully achieved.

3. After comparative consideration, the upper and lower bounds of each parameter component shrink faster by DE-FD-OEF algorithm with directional expansion compared with GE-FD-OEF algorithm, and correspondingly, the convergence rate of fault identification is faster. At the same time, affected by the conservatism of set membership estimation algorithm, the upper and lower bounds of each parameter component corresponding to the two algorithms tend to coincide gradually, and the final fault estimation values are also similar.

7.5.3   *Application Case 3*

In order to verify the effectiveness of the method proposed in this chapter to solve the linear discrete state space state estimation problem, the following uses the wind turbine pitch subsystem as a simulation example for analysis.

In the wind turbine system, the pitch subsystem is an important part of controlling the blade and pitch angle, as shown in Fig. 7.22. Its structural model is

$$\begin{bmatrix} \dot{\beta} \\ \dot{\beta}_{a} \end{bmatrix} = \begin{bmatrix} 0 & 1 \\ -\omega_{n}^{2} & -2\zeta\omega_{n} \end{bmatrix} \begin{bmatrix} \beta \\ \beta_{a} \end{bmatrix} + \begin{bmatrix} 0 \\ \omega_{n}^{2} \end{bmatrix} \beta_{r}. \tag{7.52}$$

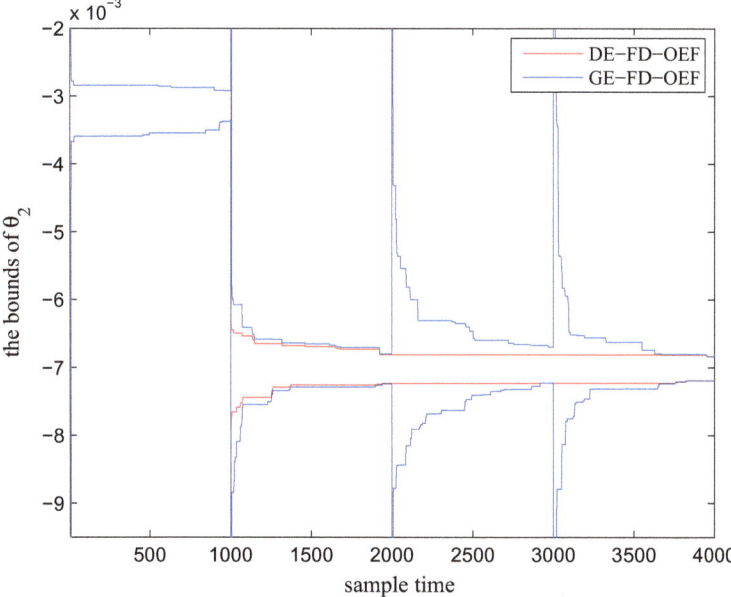

Fig. 7.18 Upper ($\theta_{x_2}^+$) and lower ($\theta_{x_2}^-$) bounds of θ_2 via different algorithms

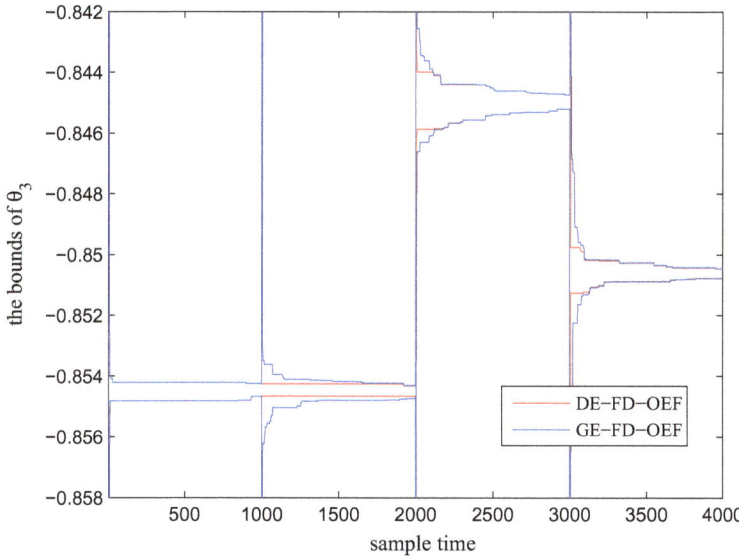

Fig. 7.19 Upper ($\theta_{x_3}^+$) and lower ($\theta_{x_3}^-$) bounds of θ_3 via different algorithms

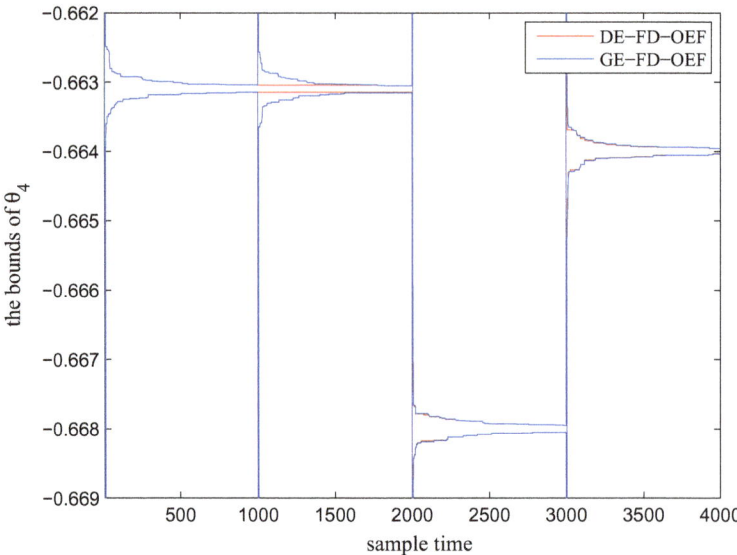

Fig. 7.20 Upper $(\theta_{x_4}^+)$ and lower $(\theta_{x_4}^-)$ bounds of θ_4 via different algorithms

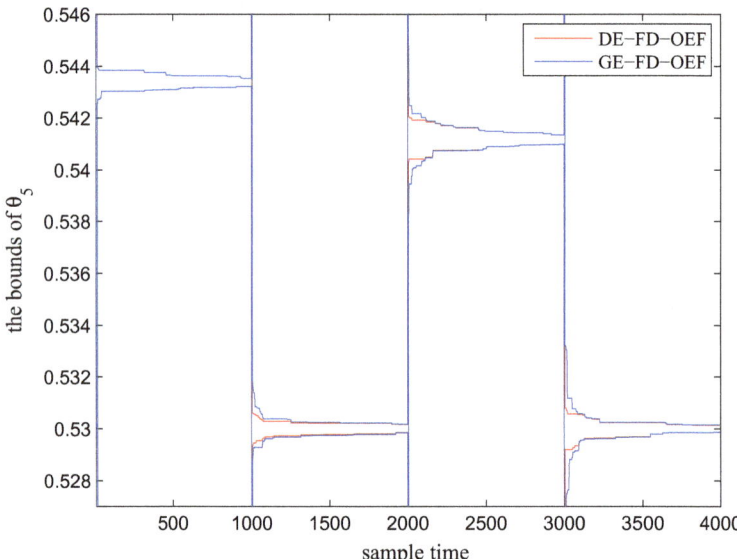

Fig. 7.21 Upper $(\theta_{x_5}^+)$ and lower $(\theta_{x_5}^-)$ bounds of θ_5 via different algorithms

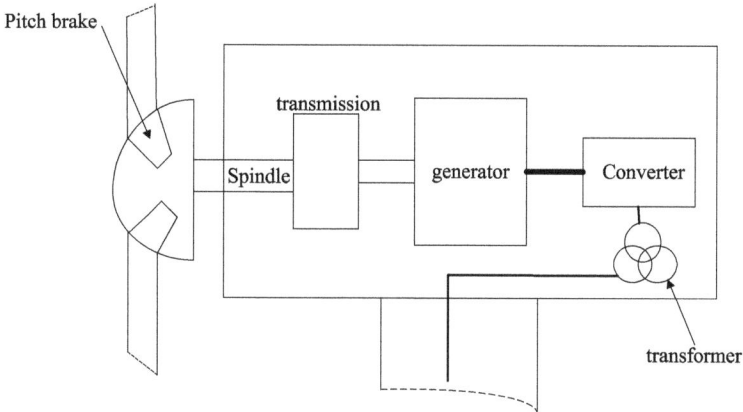

Fig. 7.22 Wind turbine system

Among them, β and β_a are the pitch angle and angular velocity, respectively, and β_r is the pitch reference value, $\zeta = 0.6$ and $\omega_n = 11.11$ rad/s are the natural frequency and damping coefficient of the system, respectively. Set state $|w(k)| = \max_i |w_i(k)| \leq \delta$, $|v(k)| = \max_i |v_i(k)| \leq \delta$, δ is the standard deviation of the sensor, take $\delta = 0.025$, sampling time $Ts = 1$ s. Discretizing the system, where

$$A = \begin{bmatrix} 0.9941 & 0.0093 \\ -1.1532 & 0.8695 \end{bmatrix}, B = \begin{bmatrix} 0.0059 \\ 1.1532 \end{bmatrix}.$$

Set $C = \begin{bmatrix} 1 & 0 \\ 0 & 1 \end{bmatrix}$.

Set initial state $x(0) = [0 \ 0]^T$, $p(0) = [0 \ 0]^T$, $H(0) = 0.6I_2$, input $u_k = 20$, system disturbance and measurement noise respectively satisfy $\|w_k\| \leq 0.025$, $\|v_k\| \leq 0.025$. The simulation results are shown in Figs. 7.23, 7.24 and 7.25.

Figure 7.23 shows the state change curve and the recursive evolution diagram of the state feasible set. It can be seen from Fig. 7.4 that the state truth value is always wrapped in the state feasible set. At the same time, Fig. 7.23 shows the zonotope obtained by the update method of literature [6], during the wrapping process, the upper and lower bounds of the zonotope did not change. It can be seen that the method proposed in this chapter always selects the optimal value closer to the true value, making the estimation interval smaller and closer to the true value.

Figures 7.24 and 7.25 show the change curves of the two states compared with literature [7], literature [6] and literature [8] algorithm, which can better demonstrate the superiority of the method in this chapter and low conservatism. It can be seen from the figure that the true value of the two states is always within the estimation interval, although the upper and lower bounds of the estimation results of each method can contain the true value. However, the algorithm proposed in this chapter

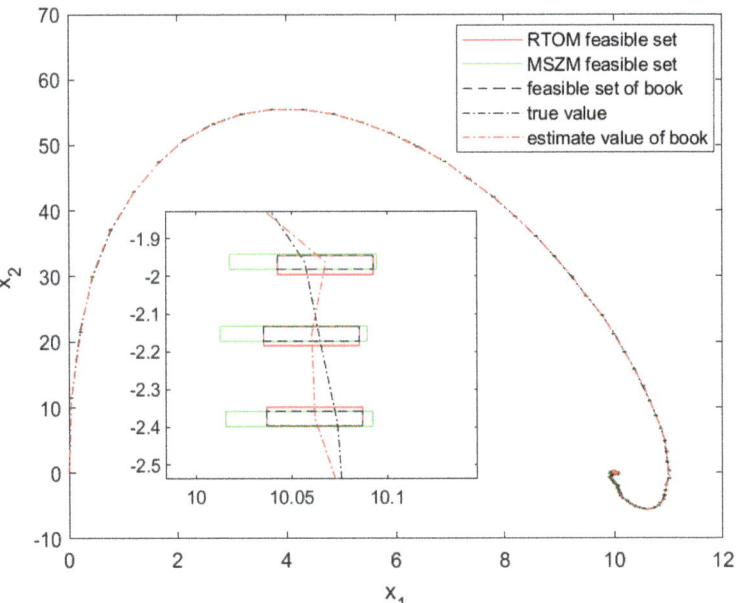

Fig. 7.23 Recursive evolution in the state feasible set estimation process

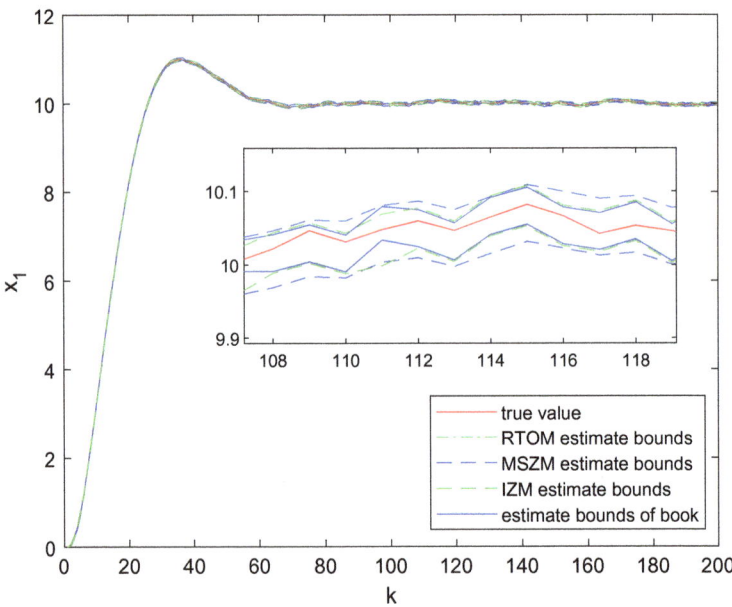

Fig. 7.24 State x_1 estimate result comparison

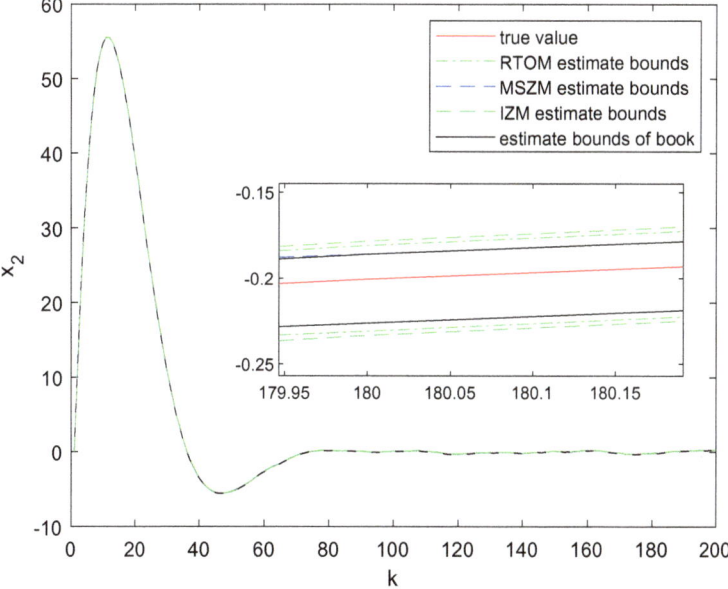

Fig. 7.25 State x_2 estimate result comparison

always chooses the best solution to estimate, and the upper and lower bounds of the estimation are closer to the true value, which reduces the conservativeness of the algorithm and makes the estimation result more accurate.

7.5.4 Application Case 4

In order to further verify the method proposed in this chapter, the four-capacity water tank system is used to show the effectiveness and superiority of the proposed method.

The mathematical model of the four-capacity water tank can be expressed as

$$
\begin{cases}
\dot{h}_1(t) = -\dfrac{a_1}{A_1}\sqrt{2\,gh_1(t)} + \dfrac{a_3}{A_1}\sqrt{2\,gh_3(t)} + \dfrac{\gamma_1 k_1}{A_1}v_1(t), \\[2mm]
\dot{h}_2(t) = -\dfrac{a_2}{A_2}\sqrt{2\,gh_2(t)} + \dfrac{a_4}{A_2}\sqrt{2\,gh_4(t)} + \dfrac{\gamma_2 k_2}{A_2}v_2(t), \\[2mm]
\dot{h}_3(t) = -\dfrac{a_3}{A_3}\sqrt{2\,gh_3(t)} + \dfrac{(1-\gamma_2)\,k_2}{A_3}v_2(t), \\[2mm]
\dot{h}_4(t) = -\dfrac{a_4}{A_4}\sqrt{2\,gh_4(t)} + \dfrac{(1-\gamma_1)\,k_1}{A_4}v_1(t).
\end{cases}
\tag{7.53}
$$

Table 7.6 Model parameters

Symbols	Parameter
$h_i(t)$	The water level of the water tank i at the moment of t
$v_i(t)$	The input voltage of the water pump i at the time of t
A_i	Cross-sectional area of the water tank i
a_i	The cross-sectional area of the outlet of the water tank i
k_i	Pump i proportional coefficient
γ_i	Valve i ratio coefficient
g	Acceleration of gravity

Among them, the model parameters of the four-capacity water tank and their physical meanings are given in Table 7.6.

The parameter values of the four-capacity water tank are respectively set as $\gamma_1 = 0.665$, $\gamma_2 = 0.566$, $k_1 = k_2 = 3.35 \, \mathrm{cm^3/Vs}$, $A_1 = A_3 = 28 \, \mathrm{cm^2}$, $A_2 = A_4 = 32 \, \mathrm{cm^2}$, $a_1 = a_3 = 0.071 \, \mathrm{cm^2}$, $a_2 = a_4 = 0.057 \, \mathrm{cm^2}$, $g = 9.81 \, \mathrm{cm/s^2}$.

Before applying the method proposed in this chapter, first use Taylor expansion to linearize the nonlinear model around the operating point. The linearized state space equation is:

$$
\begin{cases}
\Delta \dot{h}_1(t) = -\dfrac{a_1}{A_1}\sqrt{\dfrac{g}{2\bar{h}_1}}\Delta h_1(t) + \dfrac{a_3}{A_1}\sqrt{\dfrac{g}{2\bar{h}_3}}\Delta h_3(t) \\
\qquad\quad + \dfrac{\gamma_1 k_1}{A_1}\Delta v_1(t), \\[2mm]
\Delta \dot{h}_2(t) = -\dfrac{a_2}{A_2}\sqrt{\dfrac{g}{2\bar{h}_2}}\Delta h_2(t) + \dfrac{a_4}{A_2}\sqrt{\dfrac{g}{2\bar{h}_4}}\Delta h_4(t) \\
\qquad\quad + \dfrac{\gamma_2 k_2}{A_2}\Delta v_2(t), \\[2mm]
\Delta \dot{h}_3(t) = -\dfrac{a_3}{A_3}\sqrt{\dfrac{g}{2\bar{h}_3}}\Delta h_3(t) + \dfrac{(1-\gamma_2)\,k_2}{A_3}\Delta v_2(t), \\[2mm]
\Delta \dot{h}_4(t) = -\dfrac{a_4}{A_4}\sqrt{\dfrac{g}{2\bar{h}_4}}\Delta h_4(t) + \dfrac{(1-\gamma_1)\,k_1}{A_4}\Delta v_1(t).
\end{cases}
\tag{7.54}
$$

where, $\Delta h_i(t) = h_i(t) - \bar{h}_i$, $\bar{h}_i$ is the water level when the water tank i is in a stable state. $\Delta v_i(t) = v_i(t) - \bar{v}_i$, $\bar{v}_i$ is the voltage when the pump i is in a stable state. By setting the state parameters, input and output:

$$
x(t) = [\Delta h_1(t) \quad \Delta h_2(t) \quad \Delta h_3(t) \quad \Delta h_4(t)]^{\mathrm{T}} \, (\mathrm{cm}),
$$

$$u(t) = \begin{bmatrix} \Delta v_1(t) \\ \Delta v_2(t) \end{bmatrix} (V), \, y(t) = \begin{bmatrix} k_c \Delta h_1(t) \\ k_c \Delta h_2(t) \end{bmatrix} (V).$$

We can obtain the dynamic system of four-capacity water tank, as shown below:

$$\begin{cases} \dot{x}(t) = \begin{bmatrix} -m_1 & 0 & \frac{A_3 m_3}{A_1} & 0 \\ 0 & -m_2 & 0 & \frac{A_4 m_4}{A_2} \\ 0 & 0 & -m_3 & 0 \\ 0 & 0 & 0 & -m_4 \end{bmatrix} x(t) \\ \qquad + \begin{bmatrix} \frac{\gamma_1 k_1}{A_1} & 0 \\ 0 & \frac{\gamma_2 k_2}{A_2} \\ 0 & \frac{(1-\gamma_2)k_2}{A_3} \\ \frac{(1-\gamma_1)k_1}{A_4} & 0 \end{bmatrix} u(t), \\ y(t) = \begin{bmatrix} k_c & 0 & 0 & 0 \\ 0 & k_c & 0 & 0 \end{bmatrix} x(t). \end{cases} \qquad (7.55)$$

where, $m_i = \frac{a_i}{A_i} \sqrt{\frac{g}{2h_i}}$, $k_c = 0.5$ V/cm is the output voltage scale factor. Using the Euler discretization method and setting the sampling time as $Ts = 1$ s, the system dynamics of the four-tank water tank system can be formulated into the form of discrete state space, and the discretized parameters are:

$$A = \begin{bmatrix} 0.9841 & 0 & 0.0419 & 0 \\ 0 & 0.9888 & 0 & 0.0294 \\ 0 & 0 & 0.9581 & 0 \\ 0 & 0 & 0 & 0.9706 \end{bmatrix},$$

$$B = \begin{bmatrix} 0.0796 & 0 \\ 0 & 0.0.0593 \\ 0 & 0.0519 \\ 0.0351 & 0 \end{bmatrix},$$

$$C = \begin{bmatrix} 0.5 & 0 & 0 & 0 \\ 0 & 0.5 & 0 & 0 \end{bmatrix}.$$

Set the initial state $x(0) = [0.1 \quad 0.1 \quad 0.1 \quad 0.1]^T$, $p(0) = [0 \quad 0 \quad 0 \quad 0]^T$, $H(0) = 0.6 I_4$, enter $u_k = [8 \quad 8]^T$, system disturbance and measurement noise meet $\|w_k\| \le 0.2$, $\|v_k\| \le 0.2$. The simulation results are shown in Figs. 7.26, 7.27 and 7.28.

Figure 7.26 shows the plan view of the state feasible set of states x_1 and x_2. In order to clearly show the change of the state feasible set, Fig. 7.26 chooses to draw the state feasible set every 8 steps. It can be seen from the figure that the feasible set of the state can always follow the change of the state. The method proposed in this

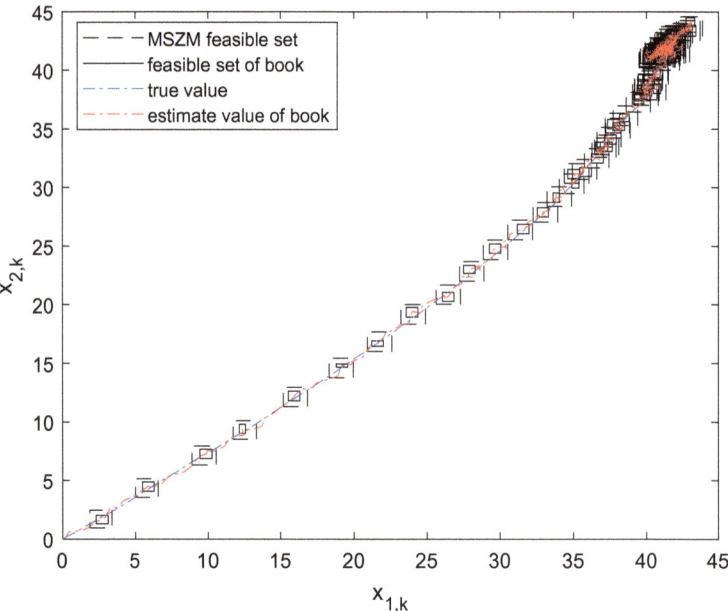

Fig. 7.26 Recursive evolution in the state feasible set estimation process

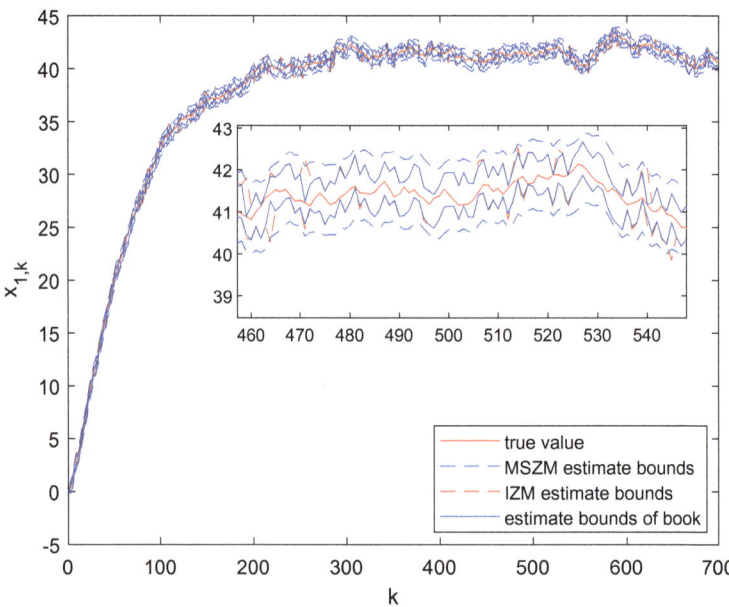

Fig. 7.27 State x_1 estimate result comparison

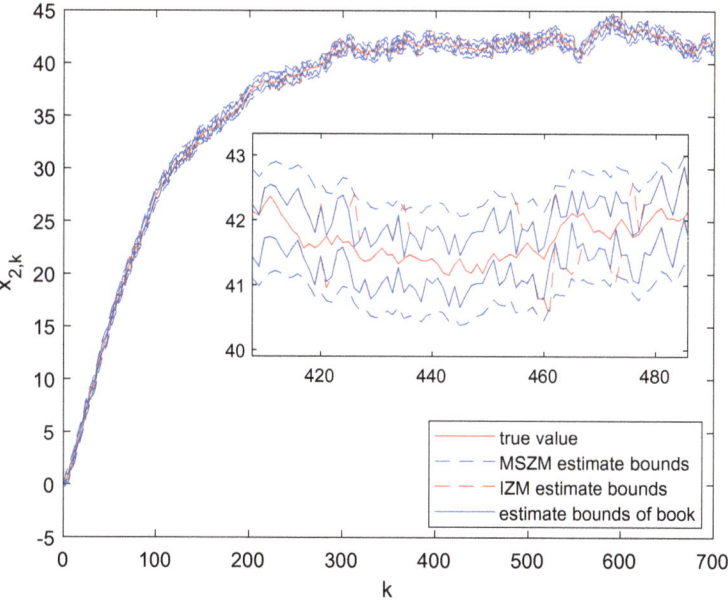

Fig. 7.28 State x_2 estimate result comparison

chapter shows good results in both the area of the feasible set of the state and the upper and lower bounds of the feasible set.

Figures 7.27 and 7.28 respectively show the change curves of the states x_1 and x_2. It can be seen from the figure that the true value of the state is always wrapped within the upper and lower bounds of the estimate. The true value of the upper and lower bounds of the method proposed in this chapter is closer than the literature [6] and literature [8]. It can be seen that the method proposed in this chapter is less conservative in solving the state estimation of the discrete-time state space system, which fully demonstrates the accuracy and effectiveness of the algorithm.

7.6 Concluding Remarks

A directional expansion based fault diagnosis algorithm using the orthotopic and ellipsoidal filtering (DE-FD-OEF) for a system with unknown but bounded noise is proposed in this chapter. The proposed fault diagnosis method uses both ellipsoidal set and orthotopic set in set membership filtering, which reduces the computational complexity and ensures the monotonic convergence of parameter bounds. More importantly, the feasible set of parameter vector is expanded only in the faulty directions by applying the directional expansion method in the DE-FD-OEF algorithm, compared to the general global expanded fault diagnosis algorithm which expands the feasible set of parameter vector in all directions without considering the faulty directions, it has the advantages of faster convergence rate and shorter fault identification time. In the process of fault diagnosis using the proposed algorithm, the system's faults are detected by performing a consistency test for the ellipsoidal set and strip, and the faults are isolated based on the state of the test sets. On this basis, the fault identification is carried out by resetting the orthotopic set directionally. The effectiveness and practicability of the proposed algorithm are verified via a numerical simulation and two case simulations in this chapter.

The proposed fault diagnosis algorithm can be applied to the fault diagnosis of other systems with unknown but bounded noise. Moreover, it can further extend the directional expansion method to the field of set membership filtering fault diagnosis in the form of other feasible parameter sets, such as the interval set [9] and the zonotopic set [10], and to deal with the control problem of some nonlinear systems, i.e., the switched nonlinear systems [11–13].

In this chapter, based on orthotopic and zonotopic filtering, combined with linear programming to solve the state parameters, a system state estimation method based on orthotopic double filtering is proposed. Under the condition of given bounded noise, it depends on the prediction of orthotopic discrete constraints, measurement constraints and minimum zonotopic constraints, In this chapter, we propose a system state estimation method based on orthotopic double filtering, which uses the spatial movement of orthotope to describe the estimation process. At the same time, in the calculation process, we get the compact orthotope by solving a finite number of linear programming problems, which is simple and low conservative, In this chapter, the pitch subsystem and four tank multi-input multi-output system of wind turbine are simulated respectively, and the feasibility and effectiveness of this method to estimate the state point can be seen directly. The research method in this chapter can be extended to the research fields of system identification and fault diagnosis.

References

1. V. Reppa, A. Tzes, Fault detection and diagnosis based on parameter set estimation. IET Control Theory Appl. **5**(1), 69–83 (2011)
2. B. Li, Z.M. Zeng, N.C. Beaulieu, et al., Wireless energy transfer beamforming with Löwner-John ellipsoidal approximation. IEEE Commun. Lett. **20**(8), 1667–1670 (2016)
3. M. Henk, Löwner-John ellipsoids. Doc. Math. 95–106 (2012)
4. Afshin Izadian, Pardis Khayyer, Parviz Famouri, Fault diagnosis of time-varying parameter systems with application in MEMS LCRs. IEEE Trans. Ind. Electron. **56**(4), 973–978 (2009)
5. Vicen Puig, Fault diagnosis and fault tolerant control using set-membership approaches: application to real case studies. Int. J. Appl. Math. Comput. Sci. **20**(4), 619–635 (2010)
6. T. Alamo, J.M. Bravo, E.F. Camacho, Guaranteed state estimatin by zonotope. Automatic **41**, 1035–1043 (2005)
7. M. Casini, A. Garulli, A. Vicino, A recursive technique for tracking the feasible parameter set in bounded error estimation. Int. J. Adapt. Control Signal Proc. **31**, 1456–1466 (2017)
8. W. Chai, X.F. Sun, J.F. Qiao, Set membership state estimation with improved zonotopic description of feasible solution set. Int. J. Rob. Nonlinear Control **23**, 1642–1654 (2013)

9. C. Zhang, D.Y. Li, Y.M. Mu et al., An interval-valued hesitant fuzzy multigranulation rough set over two universes model for steam turbine fault diagnosis. Appl. Math. Model. **42**, 693–704 (2017)
10. J.K. Scott, R. Findeisen, R.D. Braatz et al., Input design for guaranteed fault diagnosis using zonotopes. Automatica **50**(6), 1580–1589 (2014)
11. L. Liu, Y.J. Liu, D.P. Li, et al., Barrier Lyapunov function based adaptive fuzzy FTC for switched systems and its applications to resistance inductance capacitance circuit system. IEEE. Trans. Cybern. 1–12 (2019)
12. L. Liu, Y.J. Liu, S.C. Tong, et al., Integral barrier Lyapunov function based adaptive control for switched nonlinear systems. Sci. China-Inf. Sci. **63**(3), 132203:1–132203:14 (2020)
13. L. Tang, D. Ma, J. Zhao, Adaptive neural control for switched non-linear systems with multiple tracking error constraints. IET Signal Process **13**(3), 330–337 (2019)

Chapter 8
Summary

In modern society, with the rapid development of science and technology, people's requirements for system capability and control technologies are also increasing. The investment of the system and the scale of automation equipment are becoming larger and larger, and the complexity is also getting higher and higher. Once the system fails, it may cause huge noise loss and casualties, as well as environmental pollution. Therefore, it is extremely important to ensure the safety and reliability of the system. Fault diagnosis technology is one of the methods to improve the reliability and safety of the system and reduce the loss. The main research of fault diagnosis technology includes fault detection, fault isolation and fault identification, namely judging whether the fault occurs, the time when the fault occurs, the location where the fault occurs, and the size of the fault. Therefore, the research of fault diagnosis technology is very important.

After more than 40 years of development, the theory and technology of fault diagnosis have made considerable development and great progress. At this stage, a variety of mature fault diagnosis methods have been proposed. The fault diagnosis method based on filter integrates the idea of filter into the fault diagnosis, and effectively filters out all kinds of noises of the system. It is a fault diagnosis method with high accuracy and good reliability. However, the traditional fault diagnosis method based on filtering usually requires the process interference and measurement noise of the system to meet the specific distribution requirements, which is obviously contrary to the actual system interference and noise situation, so the solution of this problem is widely concerned by scholars.

The filter fault diagnosis method based on set-membership estimation is a reliable and effective fault diagnosis method without the system interference and noise meeting the requirements of specific distribution law. Compared with the traditional filtering based fault diagnosis method, it has obvious advantages in universality, practicability and robustness. Therefore, the filtering fault diagnosis method based on set-membership estimation has gradually become a research hot-spot at home and abroad in recent years.

Z. Wang et al., *Advances in Fault Detection and Diagnosis Using Filtering Analysis*, https://doi.org/10.1007/978-981-16-5959-1_8

In this book, based on the research of fault diagnosis method based on traditional filtering, the set-membership filter fault diagnosis method based on ellipsoid,
polytope, interval and polyhedron is studied. At the same time, according to the
characteristics of different space shape and the specific problems at this stage, innovative research is carried out. In addition, in a deeper level, this book proposes two
fault diagnosis methods based on composite set-membership filtering, which effectively combines the advantages of two or more space shapes. Compared with the set-
membership filtering fault diagnosis method with single space shape, this method
has obvious advantages in fault detection speed and fault recognition accuracy.

The main research results obtained in this paper are summarized as follows:

1. Aiming at the problem of parameter identification and fault diagnosis of power
   converter, a fault diagnosis method based on Inverse Kalman filtering is proposed. The simulation results show that the fault diagnosis method based on
   filtering can accurately identify the parameters of power converter components
   and diagnose system faults. Compared with Recursive Least Squares algorithm,
   it has the advantages of high accuracy and strong adaptability.
2. In view of the disadvantages of repeated calculation and low data utilization
   rate of the weight ellipsoid parameter estimation algorithm, a finite data window parameter estimation algorithm for weight ellipsoid is proposed. The algorithm uses rolling finite data window, which can reduce the calculation amount
   and improve the data utilization rate. The accuracy of parameter estimation is
   improved. The change of parameter estimation error in different noise interval
   and data window length without fault is discussed by simulation. Considering the
   parameter estimation algorithm of finite data window based on weight ellipsoid,
   a fault diagnosis algorithm of finite data window filter based on weight ellipsoid
   is proposed. The fault detection filter is designed by detecting whether the feasible parameter set is empty or not. The fault identification filter is designed by
   model matching method. The feasibility of the fault diagnosis method is verified
   by experimental simulation.
3. Two fault diagnosis methods based on polytopic set-membership filtering are
   proposed for linear discrete-time systems with unknown but bounded noise.
   Firstly, a recursive average triangulation modeling algorithm is proposed. In this
   method, the unknown noise items are arranged in a strip space at each sampling time, and the triangulation is given under the minimum variance criterion.
   Compared with the traditional triangulation method, the objective space narrows
   down in each recursive step and the feasible set is smaller than the traditional
   triangulation method. In the given algorithm, the computational complexity is
   low, the accuracy is good, and it is more suitable for the actual production. In
   addiction, a multi-objective linear-programming-based four-judgment modeling
   algorithm is proposed for unknown but bounded noise system. Because there is
   no prior knowledge about the bounded noise term, during each recursive step, the
   noise signal is warped in a strip and the hyperplanes can be obtained by samples
   of input and output signals. The feasible parameter set of a linear discrete-time
   system with bounded noise, viewed as a convex polytope, is transformed into a

polyhedral cone with increasing parameter dimension. One of the vertices of the polyhedral cone is the origin, and the polyhedral vertices can be calculated when the polyhedral cone edge vectors are determined. Moreover, by adopting the multi-objective linear programming idea, a four-judgment modeling algorithm is proposed for linear discrete-time systems. The given simulations illustrate the feasibility and effectiveness of the given algorithm.

4. A vector set inversion interval filtering based actuator fault observer design method is proposed for linear time-invariant system with unknown but bounded disturbance and noise. By considering actuator fault as augmented state vector, an augmented system equivalent to the original system is constructed. Then design a fault observer based on the augmented system to obtain the interval estimation of actuator fault. At the same time, the set inversion is carried out by using the multi-time measurement outputs. Inversion contraction interval can be used to improve the observer estimation interval by intersecting both intervals, which reduces the wrapping effect of interval computation. And the simulations verify the effectiveness and practicability of the proposed method.

5. In the research of fault diagnosis method based on polytopic set-membership filter, combined with the idea of linear programming, a filter fault diagnosis method based on orthotopic linear programming is proposed for linear discrete system with unknown but bounded noise. This method uses linear programming equation to express the constraints in the process of recursive operation, which has high feasibility and practicability in system fault diagnosis. On this basis, for the time-varying parameter system with unknown but bounded noise, a parameter identification method based on orthotopic spatial expansion filter is proposed. The bounded error method is used to model the measurement noise and parameter variation process, and the orthotopic volume is expanded by the optimized expansion coefficient, so that the orthotope contains the changed parameter values, the expansion coefficient equation is constructed by the time invariant parameter system constraints, and the first k steps for all the expansion coefficient values can be solved by the linear programming method. Select the maximum value as the final expansion coefficient and use the expansion coefficient to update the variable parameter orthotopic constraints, and solve the minimum and minimum values of each parameter to obtain the most compact orthotope of the feasible domain for wrapping the parameters. Further, in order to reduce the computational complexity of the fault diagnosis method based on orthotopic set-membership filter, a filter fault diagnosis method based on spatial dimension reduction and directional extension of parameters is proposed, which effectively reduces the computational complexity in the process of fault diagnosis.

6. In order to improve the speed and accuracy of fault diagnosis method and improve the performance of fault diagnosis, multiple set space shapes can be used to package the states or parameters of the system. This book proposed two fault diagnosis methods based on composite set-membership filter. Compared with the set-membership filter fault diagnosis method based on single space shape, it can maximize the advantages of each set space shape.

To sum up, this book mainly studies the fault diagnosis method based on filtering, and makes a supplement to the fault diagnosis method based on filtering, but there are still many problems to be further studied and explored.

1. This book mainly studies linear systems, and the study of nonlinear systems is limited to simple nonlinear systems. For other more complex nonlinear systems, such as Wiener nonlinear systems, Hammerstein-Wiener systems and Error-in-variables systems, more complex nonlinear systems need to be further studied.
2. The proposed filtering based fault diagnosis method has certain advantages in detection speed compared with the traditional residual method under probability conditions, but there are still problems such as difficulty in fault identification with similar parameters. Therefore, how to continue to improve the proposed method and set the excitation signal to further improve the accuracy and efficiency of fault identification is also the next research direction.

Lightning Source UK Ltd.
Milton Keynes UK
UKHW020728011122
411444UK00002B/43